MECHANISMS IN THE CHAIN OF SAFETY

Mechanisms in the Chain of Safety

Research and Operational Experiences in Aviation Psychology

Edited by

ALEX DE VOOGT
American Museum of Natural History, USA

and

TERESA D'OLIVEIRA
Instituto Superior de Psicologia Aplicada, Portugal

CRC Press
Taylor & Francis Group
Boca Raton London New York

CRC Press is an imprint of the
Taylor & Francis Group, an **informa** business

CRC Press
Taylor & Francis Group
6000 Broken Sound Parkway NW, Suite 300
Boca Raton, FL 33487-2742

First issued in paperback 2017

© 2012 by Alex de Voogt and Teresa C. D'Oliveira
CRC Press is an imprint of Taylor & Francis Group, an Informa business

No claim to original U.S. Government works

Version Date: 20160226

ISBN 13: 978-1-138-07225-1 (pbk)
ISBN 13: 978-1-4094-1254-0 (hbk)

Visit the Taylor & Francis Web site at
http://www.taylorandfrancis.com

and the CRC Press Web site at
http://www.crcpress.com

Contents

List of Figures

List of Tables

About the Editors

Alex de Voogt has a Master's in Linguistics (Leiden University, 1992), an MBA (Rotterdam School of Management, 1998) and a PhD in Cognitive Psychology (Leiden University, 1995). He is affiliated with the Work & Organizational Psychology Program at Maastricht University and works as an Assistant Curator at the American Museum of Natural History in New York. Alex de Voogt has a commercial pilot license for helicopters and was responsible for the acquisition and development of a helicopter simulator at Maastricht University. He has been a Member of the Royal Aeronautical Society since 2007.

Teresa C. D'Oliveira has a first degree in Applied Psychology – Social and Organizational Psychology (Instituto Superior de Psicologia Aplicada, 1992), a Masters in Organizational Behavior (Instituto Superior de Psicologia Aplicada, 1995) and a PhD in Applied Psychology (Cranfield University, England, 2000). Dr. D'Oliveira is an Assistant Professor at ISPA-Instituto Universitário de Ciências Psicológicas Sociais e da Vida where she is the Scientific Coordinator for the Organizational Psychology domain and where she lectures in Organizational Psychology and Human Factors. She was a board member of the European Association for Aviation Psychology (EAAP) where she was responsible for research and development activities in the field. Dr. D'Oliveira was involved in the organization and scientific committees of the 2004, 2006 and 2008 EAAP conferences.

List of Contributors

Nadine Bienefeld, Federal Institute of Technology, Switzerland

Kate Branford, Dédale Asia Pacific, Melbourne, Australia

Ernst Burggraaff, Air Traffic Control the Netherlands (LVNL), Amsterdam, The Netherlands

Chian-Fang G. Cherng, Department of Health Psychology, Chang Jung Christian University, Taiwan, ROC.

Teresa C. D'Oliveira, ISPA – Instituto Universitário, Lisbon, Portugal

Matt Ebbatson, Cranfield University, United Kingdom

Daniela Gundert, German Aerospace Center DLR, Hamburg, Germany

Ruth Haeusler, Zurich University of Applied Sciences, Switzerland

Don Harris, Cranfield University, United Kingdom

Brenton Hayward, Dédale Asia Pacific, Melbourne, Australia

Ernst Hermann, University of Basle, Switzerland

John Huddlestone, Cranfield University, United Kingdom

K. Wolfgang Kallus, Karl-Franzens Universität Graz, Austria

Andrew Lowe, Dédale Asia Pacific, Melbourne, Australia

Esther Oprins, TNO Defense, Safety & Security, Soesterberg, The Netherlands

Viktor Oubaid, German Aerospace Center DLR, Hamburg, Germany

Renee M. Petrilli, University of South Australia, Australia

Robert A. Roe, Maastricht University, Department of Organization & Strategy, Maastricht, The Netherlands

Rodney Sears, Cranfield University, United Kingdom

Norbert Semmer, University of Berne, Switzerland

Jian Shiu, MD, Civil Medical Center, Civil Aeronautics Administration, Taiwan, ROC

Gemma Stański-Pacis, Maastricht University, The Netherlands

Matthew J.W. Thomas, University of South Australia, Australia

Alex J. Uyttendaele, Eindhoven University of Technology, The Netherlands

Alex de Voogt, American Museum of Natural History, New York, USA

Te-Sheng Wen, Department of Holistic Wellness, Mingdao University, Taiwan, ROC

Frank Zinn, German Aerospace Center DLR, Hamburg, Germany

Acknowledgements

This publication was made possible through the generous support of the European Association of Aviation Psychology (EAAP). We owe particular thanks to the board of the EAAP, the members of EAAP and the organizers of the EAAP conferences who have facilitated a much-appreciated platform for research in Aviation Psychology. In addition, we wish to thank Connie Dickmeyer and Jennifer Steffey for their generous help and suggestions in the preparation of this volume.

Introduction:
Mechanisms in the Chain of Safety

Alex de Voogt

American Museum of Natural History, USA

Aviation psychologists have studied at least three main themes in the last one hundred years (Koonce, 1984). The First World War was the background for the development of personnel selection methods that countered pilot attrition rates in the early years of military aviation. Selection methods that were originally developed for aviation have since been widely adopted by other domains. After the Second World War, a new line of research emerged that was concerned with safety, in particular the role of the pilot in the safe operation of an aircraft. Accident statistics, safety management, human/machine interaction and human error are just a few topics that were advanced as a consequence. Again, the results of these efforts were influential elsewhere and aviation remained a leading industry where ideas on safety have been expanded and investigated. In the 1990s, a third line of research within aviation psychology that studies the interaction of people within a team became known as "crew resource management". Focus on human interaction, as related to performance, also received major attention in the aviation industry. Despite their successive origins, all three themes have continued to develop within aviation psychology and have not lost their attraction for future research. Selection methods are continuously studied, particularly in the area of air traffic control. Theories on safety are still evolving and many aspects of aviation safety are in continuous need of attention. In addition, research on crew resource management is still gaining momentum.

This volume includes these main themes within aviation psychology but not as separate topics. Over time these studies have become integrated and part of a single focus. For instance, Helmreich, Merritt and Wilhelm (1999) provide an error management approach that defines behavioral strategies taught in crew resource management. Non-technical skills have become part of assessment and training (Flin and Martin 2001) so that crew resource management is not separated from selection, assessment, safety or error management.

The main focus of aviation psychology could be summarized as "safety" since all processes studied by aviation psychologists are in some way contributing to safety in aviation. However, safety has become a matter of course rather than an option, a point of departure rather than a destination. The separate themes of research are links in a continuous chain that add up to "safety" but their concentration is on understanding the connections that best achieve this enduring goal.

The chain of safety consists of mechanisms or processes that determine its effectiveness. They are highlighted as part of this volume on aviation psychology.

The Mechanisms

Three types of mechanisms show the entanglement of selection, safety and crew resource management research.

Input mechanisms refer to the beginning of the chain—selection. Pilots or air traffic controllers enter a process of learning and training, assessment and evaluation. Learning processes feed into selection and crew performance, and, as it has been researched in crew resource management, eventually become part of individual performance, so that input mechanisms ideally integrate the results from an entire chain.

Coping mechanisms are a behavioral tool used by pilots and crew to offset or overcome stress and adversity. Research on coping mechanisms has helped to understand pilot error and challenges in crew resource management. Stress resistance has been part of selection methods and its understanding has much to offer to improve training in the aviation sector.

Control mechanisms refer to the environment in which people operate, including the organization, the safety systems and the safety climate in which people perform. These mechanisms are created to control safety by developing rules and regulations, equipment and training, as well as reporting systems that help recognize problems and develop control mechanisms to counteract them. They are at the end of the chain in this volume but are as entangled as the previous mechanisms with all the (sub)fields of aviation psychology.

While all these mechanisms can be linked to the main historical themes in aviation psychology, the purpose of this volume is to show how a better understanding of each specific mechanism, when it is studied in combination with a varied number of research methods, is particularly helpful for aviation practitioners and researchers in general. Consequently, the topics are broadly introduced in their individual chapters so that non-experts may benefit from their content. Both the broad range of approaches and the wide application of the results are, therefore, of importance.

This book is not divided into the main themes of aviation psychology. Instead, it is sectioned to reflect the rephrasing and regrouping of aviation research, as explained above, so that relationships between these themes are highlighted. Research on stress is in the company of studies on flying skill, risk assessment techniques side with a study on safety culture, and studies on selection are joined by a chapter on learning curves.

The Sections and Chapters

In the section entitled "Input Mechanisms", the first chapter discusses the integration of crew resource management studies into modern pilot training methods. The authors from the German Aerospace Center show that skills in leadership, conflict management and other interpersonal competences are required next to excellence in technical knowledge and flying skill. These advances in pilot selection are illustrated with one of the latest assessment center methods that feature empirical tests to show their effectiveness.

The chapter that follows contrasts both method and field of inspiration. Experiments on prospective memory are applied to the domain of air traffic controllers with implications for their selection. Cognitive experimental psychology, one of the first areas of psychology to be applied to aviation, still has something new to offer to this domain. The computer-based task has the simplicity necessary for selection systems and this chapter introduces the concept and understanding of prospective memory in a way that allows industry and selection professionals to benefit from its insight.

The final chapter in this section on input mechanisms concerns air traffic controllers whose assessment during their period of training is analyzed. One of the applications of this study is to adapt training to the trainee's needs and optimize pass-fail decisions during the training trajectory. The assessment contrasts the examples from the earlier chapters since it is not an added task but an added analytical tool that is central to the discussion.

From initial selection to monitored learning trajectories, input mechanisms create the basis of a chain of safety. The next section, "Coping mechanisms" concentrates on the behaviors after the initial selection and learning phases. Aviation psychologists focus on the situations that are difficult to simulate in initial training and that continue to be of concern during the career of a pilot or air traffic controller. The monitoring takes on a different guise as physiological measures enter the stage as well.

"Coping mechanisms" are diverse and this section steers away from a single approach. One may develop a new concept, such as adaptation, to improve coping skills; or, one may improve existing manual flying skills to help cope in unusual situations; or, cultural difference can be shown to influence coping skills. They all have a bearing on training and evaluation practices and are illustrated in this section, at the same time pointing out innovative directions for new research in this field.

The section starts with a group of scholars from four different academic institutions in Switzerland who developed insight in coping by looking at adaptation. They state that one cannot assume that crews trained in one type of scenario will automatically transfer the acquired skills to another scenario. Adaptation is shown to pay off, since cockpit crews who chose adaptive task strategies showed higher technical performance in the experiment set up by this Swiss team. In addition, they show the implications and possibilities for training.

While adaptation is a new concept for improving coping strategies, another four scholars, this time from the United Kingdom, focus on manual flying skills, a standard part of aviation training, that has been given less attention over time since automation seems to solve so many of the safety problems in aviation. However, they rightly point out that there are occasions when going back to basic manual control is essential or even preferable. The problems of measuring flying skill are central in their study and they argue in favor of frequency-based measures of control inputs as an additional metric dimension to performance evaluation.

A group of Taiwanese scholars and practitioners adds a study on pilot stress, the classic physiological measure of effort in tasks as complex as aviation. Their emphasis is on enhancing flight safety by establishing pilot selection criteria and reinforcing educational and training programs according to civil pilots' baseline stress levels, sources of stress, reactions to stress and coping behaviors. The differences they find between Taiwanese and non-Taiwanese pilots are particularly telling. While crew resource management studies already acknowledge the complications of multi-cultured airline companies or even airline crews, the implications for understanding stress bring a new line of research to the fore.

This section concludes with a chapter on anticipatory processes in flying. It bridges the study of coping, of which anticipatory processes are an integral part, and the study of control that shifts to an organizational perspective, for which the understanding of anticipatory processes are particularly valuable. It is shown that anticipatory processes may add aspects to accident or incident analyses as well as training, and to the analysis of situation awareness as well as selection procedures. As a synthesis of existing studies it provides a valuable overview of the insights in this complex phenomenon.

The final chapters in a section on "Control mechanisms" address control in aviation. The chapters move from error detection to accident analysis to risk assessment. They create a natural increase of scope when it comes to understanding and preventing accidents and the data required to conduct such studies. The final chapter on safety climate brings the organizational aspects into view. Each of the chapters adds original thoughts and approaches, such as new systems to collect or analyze data or new types of accident causes that can be detected in existing data sets. Their variety addresses the general concern that traditional accident analysis studies are not sufficient for the continuous improvement in safety that is expected for the future.

As can be gathered from all chapters, this volume has a human centered approach, befitting aviation psychology. Even the chapters on "control" do not focus on automation. Each chapter points toward new directions and possibilities and, certainly, new lines of research. They promise a productive future of aviation psychology.

The International Chain

Aviation Psychology is no longer limited to the military or focused on European pilots of the early wars. Since the 1950s, the United States has taken over the lead in many areas of aviation psychology research. The United States are not alone as this and many other volumes demonstrate (e.g., Lowe, Dell and Hayward 2000, Harris 2001), however, the connections or links in an academic chain are neither automatic nor superfluous. While the United States hosts a community of aviation psychologists with conferences and a journal in addition or comparable to a series of edited volumes, all European countries combined barely balance the scale on the other side of the ocean. Few aviation psychology departments have remained in Europe and many contributions from this continent come from the aviation industry rather than European universities. Today, the scholarly output in this field comes from a global community of scholars that now also include East Asia and Australia and with this expansion the balance has shifted for the better. Studies from these regions add both university and industry resources to a field that has much to gain from more international links. The issues in crew resource management are particularly relevant for the internationally mixed crews found outside the United States and Europe and the possibilities of research increase only when others and other types of research facilities join the effort. May such a chain of safety permit the field of aviation psychology to thrive through the cooperative efforts exemplified in this volume.

References

Flin, R. and Martin, L. (2001). Behavioral markers for crew resource management: a review of current practice. *International Journal of Aviation Psychology*, 11(1), 95–118.

Harris, D. (ed.) (2001). *Engineering Psychology and Cognitive Ergonomics. Vol. 6*. Aldershot: Ashgate.

Helmreich, R.L., Merritt, A.C. and Wilhelm, J.A. (1999). The evolution of crew resource management training in commercial aviation. *International Journal of Aviation Psychology*, 9(1), 19–32.

Koonce, J.M. (1984). A brief history of aviation psychology. *Human Factors*, 26(5), 499–508(10).

Lowe, A.R. and Hayward, B.J. (eds) (2000). *Aviation Resource Management. Vol. 2. Proceedings of the Fourth Australian Aviation Psychology Symposium*. Aldershot: Ashgate.

Chapter 1

GAP: Assessment of Performance in Teams – A New Attempt to Increase Validity

Viktor Oubaid, Frank Zinn and Daniela Gundert
German Aerospace Center DLR, Germany

Introduction

Safe and effective performance in aviation, for example pilot proficiency, demands not only excellent technical knowledge, but also pronounced interpersonal competence, which includes the selection and distribution of information, cooperative goal orientation and of course decision making. Moreover, skills in leadership and conflict management are required (Maschke and Rother, 2006).

Modern pilot training methods reflect these demands. "Human performance and limitations", "multi-crew coordination training" and "crew resource management training" are important subjects in the airline pilot education and licensing process.

In order to select skilled candidates for training, a psychological selection including group assessment methods is employed by many airlines and air traffic controller organizations (Goeters, 2003). These methods typically comprise group discussion, group planning and prioritization tasks. The applicant's behavior is observed, registered by the observers using paper and pencil, and lastly, rated within a given guideline framework.

With the advance of computer and information technology there are two significant reasons why these methods require updating (Huelmann and Oubaid, 2004). First, typical tasks in the working environment of pilots are located at the human-machine gateway. Second, in conventional assessment center exercises, the objectivity of behavior ratings is comparatively lower than in other psychological methods. This is due to observer errors, complex interactions, and time-absorbing behavior reporting (paper and pencil).

A further problem of those aforementioned methods is the often low interrater reliability among different observers. This is also a direct result of the varying consideration of which exhibited behavior is deemed significant for a certain dimension. Additionally, noting observations by hand draws the observer's attention away from the continuing process of interaction. This results in missing or misperceiving behavioral units, an additional negative effect on reliability.

To overcome those problems, a computer-based group test system (GAP Assessment®; Group Assessment of Performance and Behaviour; Oubaid, 2007; Oubaid, Zinn and Klein, 2008) was developed in which behavior observations made by four experts (consisting of training captains and aviation psychologists) and objective behavior measures are integrated into an overall evaluation. The basis of the multi-level observations are taxonomically derived complex scenarios in which three or four applicants gradually receive different assignments and interact with each other face-to-face as well as through their individual touch-screen monitors that are part of the GAP network (see Figure 1.1).

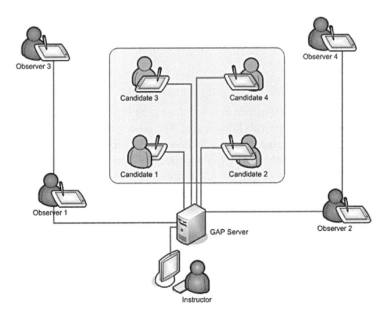

Figure 1.1   Overview of the GAP Assessment® network including four observers and four applicants

Method

As a first step, a behavioral observation model was developed that also functions as the backbone for the scenario construction. This model is based on three sources: (1) the set of basic competencies used in Lufthansa pilot training, which includes basic interpersonal, technical and procedural competencies for a safe flight accomplishment (Lufthansa, 1999). (2) The VERDI Circumplex Behavioral model for DLR pilot selection (for example, Hoeft, 2003). (3) A Fleishman job requirement analysis for airline pilots (Maschke, Goeters and Klamm, 2000) was integrated to elaborate the areas of competence. Six areas of competence could be identified: leadership, teamwork, communication, decision making, adherence to procedures, and workload management.

As a second step, individual behavioral units – the behavioral anchors – were derived to translate the areas of competence into practice. These anchors were presented to assessment-center experts (aviation psychologists, training captains) to be rated regarding their prototypic assignment to the areas of competence following the Act Frequency Approach (Buss and Craik, 1980, 1983). The behavioral anchors were then combined into behavioral subsets (GAP sets). These GAP sets are the basis of behavioral observation in GAP scenarios. In the final version, three different categories of strain symptoms (hypomotor/hypermotor, vegetative, and paralinguistic symptoms) were integrated.

During four sequences the observers use a touch-screen to assign behavioral anchors to register the behavioral units which were presented by the applicants (see Figure 1.2).

The observer's screen also includes:

*   feedback information about tasks the applicants are currently working on;
*   feedback about certain performance parameters like talk-time, matching-task, or errors made;
*   notifications about individual and group messages given by the instructors to intervene.

After each sequence, additional clinical ratings of the past performance are given by observers (for all applicants) and by the applicants (self-rating and peer-rating): rating leadership, teamwork and effectiveness. The observers rate two further dimensions: stress resistance and authenticity.

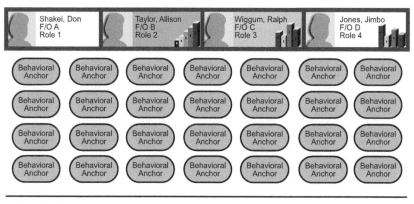

Figure 1.2   Overview of the GAP Assessment® observer screen

Typically, candidates tend to overestimate their behavior in assessment centers (Sarges, 1990, p. 529). GAP Assessment® enhances the quality of self-evaluation by implementing objective self-awareness during the self-evaluation process. According to the Self-awareness theory, people who focus their attention on themselves evaluate and compare their behavior to their internal standards and values (Duval and Wicklund, 1972). In GAP Assessment®, the self-awareness condition is assisted by presenting the candidate's photograph during self-rating. The photograph is taken during instruction time and also used for the subsequent peer-rating.

To augment the connection between the set of required competencies, the behavioral anchors and the roles played by applicants, scenarios were created on the basis of these anchors. The resulting GAP Assessment® scenario consists of four sequences. Sequences one and three contain schedule planning. Sequences two and four are conflict tasks in which the individual interests are not fully compatible to each other. In one example scenario, each candidate is one of three or four flight attendants. His or her role involves planning, which entails the re-arrangement of passengers on a flight and rotation schedules. The role also involves conflict tasks, which include group decisions about an unattractive rotation and the nomination of an executive position. The information presented on the applicant's screen contains both individual and group role details, task instructions and a set of working rules and restrictions. To estimate the applicant's cognitive workload, a simple matching task is presented during the whole scenario. It involves a pair of randomly drawn letters with a refresh-rate of five seconds (see Figure 1.3).

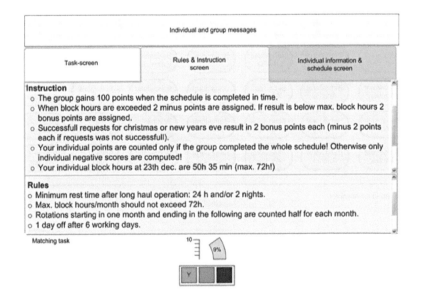

Figure 1.3   Example of the GAP Assessment® applicant's screen

Results

The analysis focussed on three aspects:

- statistics for the GAP anchor measures;
- the quality of additional dimensional post-sequence ratings (self-rating versus peer- rating and observer rating);
- correlations between GAP variables and DLR assessment variables.

The sample consisted of N = 131 applicants, n = 115 males, and n = 16 females. The ratio of males versus females reflects approximately the typical ratio of pilot applicant groups. The mean age was 20.9 years. The data was collected between May and July 2010.

The GAP competence areas did not correlate significantly with age or school grades (A-level). However, there was one exception, decision making multiplied by age (see Table 1.1).

Table 1.1   Correlations of GAP anchor scores with biographical data

	Age[1]	Sex[2]	Schoolgrades: A-level[1]
Leadership	−.15	−.04	−.02
Teamwork	−.12	.15	−.13
Communication	−.13	.06	−.06
Adherence to procedures	−.09	.11	.03
Awareness / Workload	.08	−.02	−.09
Decision making	−.18*	.01	−.02
Strain symptoms	−.09	.03	−.05
Matching task score	−.14	.01	.06

Note: N = 131, * p<.05; [1] Pearson coefficients, [2] Spearman coefficients

Table 1.2 displays the correlations between the post-hoc observer ratings with self-ratings and peer-ratings.

Table 1.2   Pearson correlations of observer ratings with self-ratings and peer-ratings

Observer ratings	Self-ratings			Peer-ratings		
	Leadership	Teamwork	Effectivity	Leadership	Teamwork	Effectivity
Leadership	.29***	−.05	.03	.53***	.19*	.25**
Teamwork	.01	−.06	−.04	.09	.18*	.05
Effectivity	.16	−.01	.03	.34***	.17	.16
Peer-ratings						
Leadership	.38***	.04	.00			
Teamwork	.11	.14	.06			
Effectivity	.20*	.08	.10			

Note: N = 131, * p<.05, ** p<.01, *** p<.001

Self-ratings and peer-ratings show a high congruence for leadership.

Table 1.3   Pearson correlations of observer's GAP anchor scores with GAP dimensional ratings

Anchor scores	Dimensional ratings				
	Leadership	Teamwork	Stress resistance	Effectivity	Authenticity
Leadership	.49***	.13	.36***	.37***	.18*
Teamwork	.12	.42***	.25**	.20*	.34***
Communication	.25**	.04	.19*	.22**	−.04
Adherence to procedures	.12	.14	.07	.14	.14
Awareness / Workload	.42***	.33***	.37***	.42***	.20*
Decision making	.46***	.31***	.34***	.42***	.19*
Strain symptoms	.21*	.11	.39***	.19*	.10
Matching task score	.14	−.02	.17*	.11	.02

Note: N = 131, * p<.05, ** p<.01, *** p<.001

The anchor-based observation results in a comparatively high accordance of judgements with the post-hoc dimensional ratings (Table 1.3). The high correlations of decision making and awareness/workload anchor scores with all dimensional ratings are remarkable and can be attributed to the highly structured GAP tasks.

In the next analysis, GAP anchor scores were compared with the judgements of assessment center experts using the VERDI Circumplex Behavioral model (Hoeft, 2003) and the DCT-Scheme (Stelling, 1999) for DLR pilot selection. The statistical analysis includes comparisons of these judgements on different levels (see Tables 1.4 and 1.5).

Table 1.4 Correlations of GAP anchor scores with VERDI dimensional ratings

	VERDI ratings			
GAP scores	**Coordination**	**Cooperation**	**Activation**	**Stress resistance**
Leadership	.40***	.16	.36***	.19*
Teamwork	.18*	−.10	.22*	.01
Communication	.31***	.03	.26**	.17
Adherence to procedures	.19*	.08	.17	.14
Awareness / Workload	−.12	.03	−.11	−.10
Decision making	.33***	.11	.30***	.16
Strain symptoms	−.20*	−.09	−.17	−.42***
Matching task score	.14	.13	.17	.18*

Note: N = 131, * p<.05, ** p<.01, *** p<.001

Table 1.5 shows the correlations of GAP anchor scores with DCT scores.

Table 1.5 Correlations of GAP anchor scores with DCT scores

	DCT scores					
GAP scores	**Work style**	**Cooperation**	**Reliability / Discipline**	**Decision making**	**Stress resistance**	**Solo score**
Leadership	.08	.02	−.03	.17	.02	.24**
Teamwork	.11	.22*	−.14	.04	−.13	.01
Communication	.19*	.13	−.03	.05	−.02	.17*
Adherence to procedures	.05	.21*	−.04	−.04	−.01	−.01
Awareness / Workload	−.10	.15	−.08	−.16	.01	−.09
Decision making	.08	−.02	.04	.12	.03	.28**
Strain symptoms	−.15	−.09	−.15	−.13	−.31***	−.03
Matching task score	.16	−.05	.02	.00	.10	.06

Note: N = 131, * p<.05, ** p<.01, *** p<.001

Most correlations are plausible and convincing. As expected, correlations between GAP leadership and VERDI coordination are highest due to their common definition. Also, the significant correlation between the matching task and VERDI stress resistance is easy to interpret, because VERDI stress resistance involves the ability to work under stress. The main unexpected result is the low correlation between VERDI cooperation and all GAP variables, namely GAP teamwork. A possible explanation is that VERDI cooperation involves more non-verbal aspects.

The correlations with the DCT are mostly low. One explanation may be that the DCT has considerably tighter interaction requirements than GAP. Another aspect is that the DCT employs only a dyadic interaction. Nevertheless, strain symptoms show an expected high negative correlation with stress resistance. The correlations between DCT solo score and some GAP scores can be interpreted from the circumstance that the solo part of the DCT involves much decision making and goal orientation, which is an integral part of the GAP leadership definition.

In a final analysis, the GAP anchor scores for accepted applicants were compared to those of the rejected applicants in the VERDI assessment to distinguish if positive applicants show better anchor scores in GAP (see Table 1.6).

Table 1.6 **Comparison of mean GAP anchor scores: accepted (n = 60) versus rejected (n = 71) applicants in the VERDI assessment**

	Accepted applicants	Rejected applicants	p
Leadership	55.84	34.93	p<.001
Teamwork	42.99	21.50	p<.01
Communication	126.25	106.72	p=.01
Adherence to procedures	−8.31	−11.47	ns p=.35
Awareness / Workload	−5.41	−26.08	p<.001
Decision making	41.36	23.55	p<.001
Strain symptoms	−16.95	−26.98	p<.05
Matching task score	.76	.73	ns p=.15

Note: N = 131; ns = not significant

The results show that the anchor scores differ significantly between accepted and rejected applicants in the expected direction.

Discussion

The empirical results show clear evidence for the reliability and validity of GAP scoring procedures. The correlations with the corresponding measures in the DLR assessment are high and significant. One possible explanation for the higher correlations between GAP ratings and VERDI ratings compared to DCT scores is that the DCT demands the interaction of only two candidates, whereas the GAP group discussion, as well as the VERDI group task, demand the interaction of four candidates (in a group situation). Further studies have already been planned to intensify activities on this topic and also to focus on the criterion validity.

The comparison of anchor scores and dimensional ratings revealed a high significance of leadership in the observation of individuals in the group task. This is unsurprising as the scenario and the actual job requirement place a high emphasis on planning skills and leadership.

The use of photos to activate the objective self-awareness condition during self-evaluation and peer-evaluation is a positive addendum to assessment center techniques. The results show that peer-ratings and self-ratings correspond highly. Additional research should focus on the experimental variation of the objective self-awareness effect.

Finally, the design of the GAP group task and the combination of group acting elements and registered task performance allows continued research into the clarification of the coherence between individual and team performance.

Summary

The development of a new and innovative form of group assessment center method (GAP Assessment®) enhances behavior observation. A set of GAP behavior ratings was created and empirically tested. The first empirical study focused on the interrater-reliability and the validity of GAP ratings. The results show a high interrater-reliability and a good validity of GAP-behavior-ratings.

References

Buss, D.M. and Craik, K.H. (1980). The frequency concept of disposition: dominance and prototypically dominant acts. *Journal of Personality*, 43, 379–92.

Buss, D.M. and Craik, K.H. (1983). The act frequency approach to personality. *Psychological Review*, 90, 105–26.

Duval, S. and Wicklund, R.A. (1972). *A Theory of Objective Self-Awareness*. New York: Academic Press.

Goeters, K.M. (2003). The validity of pilot selection: results of studies with ab-initio and direct-entry applicants. *International Summer School on Aviation Psychology*, Graz, June 30–July 06, 2003.

Hoeft, S. (2003). Basic concepts of and research findings for assessment centers. Presentation at the EAAP course: psychological evaluation and selection of aviation personnel. Hamburg, Germany, March 27–30.

Huelmann, G. and Oubaid, V. (2004). Computerized aptitude testing in aviation psychology. In: K.-M. Goeters (ed.), *Aviation Psychology: Practice and Research*. Aldershot: Ashgate.

Lufthansa (1999). *Basic Competence for Optimum Performance*. Internal publication, Lufthansa Training Department.

Maschke, P., Goeters, K.-M. and Klamm, A. (2000). Job requirements of airline pilots: results of a job analysis. In: B.J. Hayward and A.R. Lowe (eds). (2000), *Proceedings of the Fourth Australian Aviation Psychology Symposium*, Manly, March 1998. Aldershot: Ashgate.

Maschke, P. and Rother, J. (2006). Selection and training in flight operation: Civil aviation. In: *Proceedings of the 27th Conference of European Association for Aviation Psychology*, Potsdam, September 24–28 <http://www.eaap.net/get/82.pdf> (accessed September 1, 2011).

Oubaid, V. (2007). *Group Assessment of Performance and Behaviour – GAP.* DLR Project plan. DLR.

Oubaid, V., Zinn, F. and Klein, J. (2008). Selecting pilots via computer based measurement of group and team performance – development of a new assessment center method. *Proceedings of the 28th Conference of the European Association for Aviation Psychology*, Valencia, Spain, Oct. 27–31, 2008, <http://www.eaap.net/get/82.pdf> (accessed September 1, 2011).

Sarges, W. (1990). *Managementdiagnostik [Management Diagnostics]*. Göttingen: Hogrefe.

Stelling, D. (1999). *Teamarbeit in Mensch-Maschine-Systemen [Teamwork in Man-Machine-systems]*. Göttingen: Hogrefe.

Chapter 2

The Importance of Prospective Memory for the Selection of Air Traffic Controllers

Alex J. Uyttendaele
Eindhoven University of Technology, The Netherlands

Alex de Voogt
American Museum of Natural History, USA

Introduction

Air traffic controllers are responsible for the safe travel of multiple aircraft simultaneously. A timely but complex set of commands must be communicated for a safe flight operation, particularly in congested airspace. For example, in 1991 a tower controller instructed an aircraft to hold on Runway 24, intending to clear it for take-off as soon as other traffic had passed. The controller was pre-occupied with making radio transmissions and monitoring other aircraft that caused the take-off clearance to be forgotten. Instead, another plane was ordered to land on Runway 24, which resulted in a catastrophic accident (Aviation Safety Network, 1991). This human error can, at least partially, be attributed to a memory failure exacerbated by multiple tasks that needed to be performed by the responsible controller. The specific kind of memory that was deficient in this case is called prospective memory.

Prospective memory is defined as the ability to remember and execute specific delayed tasks without being prompted at the desired time of execution, in other words remembering to remember (Einstein, McDaniel, Williford, Pagan, and Dismukes, 2003; Nowinski and Dismukes, 2005) .

Prospective memory can best be explained by describing its different phases and components. The four main phases are an encoding phase, a period of delay, an executionary phase, and finally, an evaluative phase. During the encoding stage one forms the intention to perform a certain action in the future; in the example above, Air Traffic Control (ATC) intended to clear the aircraft when possible. Once the intention is there, the required action has to be remembered (retrospective component), as well as the moment at which to execute this action (prospective component). During the delay phase one may be occupied with other kinds of responsibilities lasting up to the specific time for execution. In the executive phase the action has to be performed at the right moment, in the right place and in the way prescribed; this is where ATC failed in the mentioned example. Once the action is complete one should remember to have performed the action so that it is not repeated (evaluative phase) (Brandimonte and Passolunghi, 1994).

There are two types of prospective memory: time based and event based (Kliegel, Martin, McDaniel, and Einstein, 2001; Logie, Maylor, Della Sala, and Smith, 2004; Smith, 2003). Time based, also called self cued prospective memory, is when the task has to be executed at a specific and fixed time: for example going to an appointment at ten o'clock. Event based or externally cued is the one most relevant to ATC. At the onset of a certain event one must be triggered to execute a specific command; in the above example, ATC has to clear a plane to take-off, therefore ATC must take into account that multiple prospective tasks can occur simultaneously.

There are several theories that strive to explain how prospective memory works and the implications it has on our cognitive ability. The monitoring theory suggests a constant monitoring of the environment for the appearance of the prospective cues (Einstein et al., 2005; Smith, 2003). This theory was rejected because the monitoring would strain the working memory too much, and make it impossible to perform other tasks during the delay phase (see also: Einstein et al., 2005; Smith and Bayen, 2005). Spontaneous retrieval theory states the complete opposite. There is no active monitoring and the onset of the prospective cue causes the intentions to automatically pop back into consciousness (Einstein et al., 2005; McDaniel, Guynn, Einstein, and Breneiser, 2004). This theory was also rejected because performance on tasks given in experimental conditions suffered significantly from the prospective intentions, thus showing that a certain amount of resources is needed to find the specific cues (Einstein et al., 2005; McDaniel et al., 2004). A compromise between these theories is found in the Preparatory Attentional and Memory processes theory (PAM), and it is currently the generally accepted theory. It states that resources are always needed to monitor for the onset of the cues, but the amount of resources differs. When the prospective cue is difficult to encounter, a greater amount of resources is dedicated towards finding it. This decision is made by preparatory resources that analyze the task and decide on the amount of resources needed to encounter the prospective signal (Marsh, Hicks, and Cook, 2005; Smith, 2003; Smith and Bayen, 2004; Smith and Bayen, 2005).

The experiment presented here tests whether prospective memory is a quality that is acquired over time or whether it is an innate quality. If it is acquired over time, ATC training may focus on this quality. In this case, prospective memory can be used as a performance marker over time. Whereas, if proven to be innate, it may become a deciding factor during the selection process.

Event based prospective memory tasks are relatively simple to simulate in experimental settings. Subjects are asked to respond to the appearance of the prospective cue while simultaneously performing a cover task during the delay period. The prospective task should be of the same nature as the cover task, since this is how they usually present themselves in real life (Stone, Dismukes, and Remington, 2001). Contrary to general expectations none of the experiments included in the literature found proof for a decay of performance due to an increasing delay period (Brandimonte and Passolunghi, 1994; Einstein et al.,

2003; Stone et al., 2001) and this result will also be investigated in more detail in the present experiment.

Method

A between-subject design was used containing four groups. The variable was the level of experience as an air traffic controller. In total 54 participants of different experience levels were tested: 12 newly acquired trainees, 14 air traffic controllers who had just completed the training phase, 20 highly experienced air traffic controllers with at least three year's experience in the field, and a randomly selected control group of non air traffic controllers consisting of 12 participants. The control group allows a differentiation between those who practice or have experience with prospective memory and those who do not. The control group was comprised of participants with an equal mean age to the total population of air traffic controllers. All participants were volunteers and no awards were given for participation.

Prospective memory was measured by means of a computer task.

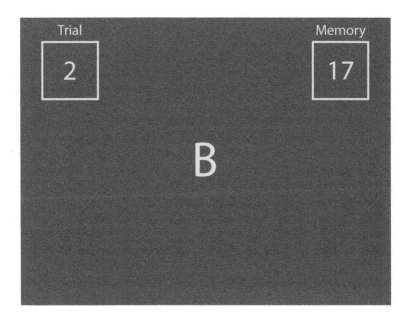

Figure 2.1   Example of cues presented on a computer screen

As a cover or filler task, subjects were asked to respond to the presentation of the letters A or B on the screen by pressing the appropriate labeled buttons. These letters were displayed for a time interval of three seconds or until a button press was recorded. Each test consisted of 50 practice trials and 200 actual trials. A counter in the top left corner continuously indicated the trial number. During 51 of the 200 trials a prospective memory signal occurred. This signal consisted of the appearance of a future trial number in the top right corner. The subjects were instructed to remember this future trial number and withhold from any response during this future trial. In other words subjects should not respond to the displayed 'A' or 'B' in the future prospective trial. These future prospective trials would always occur after one of three fixed intervals, 15 seconds (NO GO 15), 30 seconds (NO GO 30) or 45 seconds (NO GO 45), following the onset of the prospective memory signal. The test was designed in such a way that different prospective trials would overlap each other since real life situations often present coinciding prospective cues. The maximum amount of prospective cues to be remembered simultaneously was set at three. As soon as a prospective trial passed, a new cue would present itself. The dependent variables were correct responses to the presented 'A' and 'B' in the non-prospective trials (GO trials) and correctly refraining from responding in the prospective trials (NO GO trials).

Statistical Analyses

A Univariate Analysis of Variance was performed to discover whether the different experience levels (newly acquired trainees, new air traffic controllers, experienced air traffic controllers and the control group) interacted with the different delay periods (15, 30 and 45 seconds).

This analysis was followed by a One-way ANOVA that determined the differences in performance of each experience level over the different conditions (GO trials, NO GO 15, NO GO 30 and NO GO 45). If significant results were obtained, a post-hoc Bonferroni correction, was performed to assist in the interpretation of the results.

Another One-way ANOVA was conducted to measure the performance of each individual experience level over the different delay periods. This tested the findings of previous experiments that had found no significant effect on performance with an increase in delay time. Again possible significant results were submitted to a Bonferroni correction.

For all previously stated tests the rejection level for statistical significance was set at .05.

Results

Univariate Analysis of Variance

A univariate analysis of variance was conducted with the delay times and the experience levels as Fixed Factors and the different delay conditions combined as the dependant variable. This established a possible interaction between the factors and validated its use for further analysis of the data.

Experience level and delay time showed a significant interaction, $F(1,11)$ 4.76, $p < .001$.

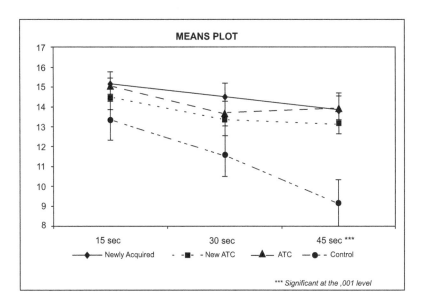

Figure 2.2   Interaction of experience levels with delay periods

Analysis of the Conditions on the Different Experience Levels

All values (means, standard deviations, f-values and p-values) of the following analyses are found in Table 2.1.

A One-way ANOVA was performed over the different experience levels with the go trials as the dependent variable. A trend to significance was found, $p = .06$. Post-hoc comparisons showed that a significant trend was found between the newly acquired trainees and the control group, $p = .06$.

Another One-way ANOVA was conducted for all the experience levels on the three delay conditions (NO GO Trials). The NO GO 15 showed no significant results. On the NO GO 30 a trend to significance was observed, $p = .09$. The

Table 2.1　　**Means, standard deviations, f-values, p-values and post hoc comparisons of the one-way ANOVAs conducted for the GO trials and the no go conditions**

	M (SD)	F (p)	Post Hoc
GO Trials	135.12 *(7.75)*	2.59 *(.06)*	Newly acquired > Control .06
NO GO 15	14.53 *(2.40)*	1.73 *(.17)*	
NO GO 30	13.24 *(3.04)*	2.24 *(.09)*	Newly acquired > Control .09
NO GO 45	12.59 *(3.44)*	7.59 *(<.001)*	Newly acquired > Control .001 New ATC > Control .006 ATC > Control < .001

Bonferroni correction revealed the trend to significance between the newly acquired trainees and the control group, $p = .09$. A significant effect was found on the NO GO 45, $p < .001$. Post-hoc comparisons revealed that the newly acquired trainees scored significantly higher than the control group at the 45 second interval, $p = .001$. New air traffic controllers also showed significantly higher results than the control group, $p = .006$. Even the experienced air traffic controllers scored significantly better than the control group, $p < .001$.

Thus all air traffic controllers, regardless of their level of experience, scored significantly better than the control group at the 45 second interval.

Analysis of Experience Levels with Different Delay Periods

A summary of all the values (means, standard deviations, f-values and p-values) of the following analysis is found in Table 2.2.

The different delay periods were analyzed using a One-way ANOVA with the experience level as the dependent variable. For the newly acquired trainees no significant result was found, $p = .41$, neither the new air traffic controllers, who produced no significant difference, $p = .32$, nor the experienced air traffic controllers, who showed no significant difference, $p = .16$. The Control group, however, did show a significant effect, $p < .05$ The post-hoc analysis showed

Table 2.2   **Means, standard deviations, f-values, p-values and post hoc comparisons of the one-way ANOVAs conducted for different delay intervals**

	M (SD)	F (p)	Post Hoc
New Trainees	14.50 *(2.40)*	.92 *(.41)*	
New ATC	13.65 *(2.39)*	1.18 *(.32)*	
ATC	14.22 *(2.31)*	1.88 *(.16)*	
Control	11.33 *(4.16)*	3.74 *(.03)*	15 Seconds > 45 Seconds .03

that this significant result was found between the 15-second and the 45-second interval, $p = .03$.

Discussion

The main objective of the experiment was to test whether prospective memory can be trained. In order to test this hypothesis, a computerized prospective memory task developed at Maastricht University was conducted with four different groups of participants. The groups differed in their experience level as air traffic controllers. Air traffic controllers are assumed to deal with prospective memory all the time during the execution of their job. The experiment tested whether prospective memory can be trained or whether it is an innate quality, and thus a relevant job criterion for selection. The experiment was conducted with a group of experienced air traffic controllers, a group of new air traffic controllers, a group who just recently started their training as air traffic controllers and a randomly assigned control group. Statistical analysis supports the notion that it is an innate quality. No matter how experienced, all ATC groups performed equally well. The Control group, however, scored significantly lower than all other ATC groups. This result was limited to the 45-second delay condition. This suggests that prospective memory is a valid selection marker for ATC. Using prospective memory as an early selection marker might help in reducing the large number of yearly drop-outs during training after having passed the selection (Oprins, Burggraaff, and van Weerdenburg, 2006).

The delay period between the onset of the prospective cue and the actual event was manipulated (15, 30 and 45 seconds). Statistical analysis rejected the previously found result that the duration of the delay has no influence on prospective memory performance (Brandimonte and Passolunghi, 1994; Einstein et al., 2003; Stone et al., 2001). The present experiment produced a significant decay of performance over an increased interval size. However, only the control group showed this decay of performance, and only between the 15 second interval and the 45 second interval. No significant differences were observed between the 15 and 30 second interval, or the 30 and 45 second interval. This suggests that small increases have no effect, but larger increases do. An experiment with even larger delay periods could show whether this finding repeats itself, or even increases with even larger delay periods.

Our results suggest that non-ATC people are hampered by an increasing interval period. Only those with a high prospective memory performance are immune from this performance loss. Previous experiments were also conducted on randomly assigned groups, thus equal to our control group. Therefore the explanation has to be found in the nature of the task itself. In most event-based prospective memory tasks the prospective cue is fixed. In Einstein's example (Einstein et al., 2003), participants were to divert from regular actions upon appearance of a red screen. Participants were thus asked to merely monitor for this one specific cue. In real life situations people respond only once to each prospective memory event. As mentioned in the theoretical section, the evaluative phase is especially meant to remember that an action has been completed and should not be repeated in the future. This theory was incorporated in the present experiment by having constantly changing prospective cues in the form of altering future trial numbers. Participants are thus required to constantly remember different prospective cues, and forget the ones that have passed. This dynamic nature might be one of the components that lead to the deteriorating performance over growing interval sizes.

Another factor that might have affected the difficulty of the present task is the requirement that subjects have to remember more than one prospective cue at the same time. At the start of the experiment new cues are presented rapidly until the memory set contains three different prospective cues. During the remainder of the task a prospective trial is always followed by the appearance of a new prospective cue. Therefore the memory-set has to be constantly refreshed, forgetting the past prospective trials and incorporating new future prospective trials. Again this simultaneous remembering and forgetting was part of the experiment to simulate real life situations.

Nevertheless, one of the previously published experiments also used a dynamic, simultaneous, event-based prospective memory task (Stone et al., 2001). Stone's experiment consisted of thirteen independent trials of seven minutes long. Each such trial contained three different overlapping prospective cues with delay times of one, three and five minutes. The prospective signal was presented in the middle of the screen for the duration of ten seconds. After only three prospective cues a trial was over and subjects had a break before starting a new trial. This break time could be used to erase the old cues from memory and prepare for the appearance of new ones. The present experiment does not include such a break period; instead new cues just keep on coming. The subjects in the present experiment are required to constantly run through the whole cycle of phases of prospective memory. As soon as they have completed the evaluative phase, they have to start with a new encoding phase again without ever stopping the cover task. The nature and presentation of the cues is another point of difference. The prospective cue is not presented centrally, but rather in the top right corner of the screen, outside the focal field. When cues are presented outside the focal field extra resources are required (Einstein et al., 2005; Kliegel, Martin, McDaniel, and Einstein, 2004; McDaniel and Einstein, 2000), causing greater strains on the working memory. The presentation of prospective cues outside of the focal field forces subjects to use more resources to monitor for the appearance of cues. Even the duration of the presentation differs drastically from Stone's: ten seconds as opposed to three seconds. Finally, Stone used larger delays of one, three and five minutes as opposed to 15, 30 and 45 seconds. Future research may assess which of these differences leads to the decreasing performance during increasing intervals.

Where the results on the delay times provide interesting material for future experiments, the main question about prospective memory has been answered. The results suggest that it is an innate quality, and thus, an appropriate early selection marker for ATC. More importantly, it suggests that prospective memory provides an initial predictor of future performance.

References

Aviation Safety Network (1991). ATC transcript US Air Flight 1493 collision–01 FEB 1991. Retrieved from <http://aviation-safety.net/investigation/cvr/transcripts/atc_us1493.php>.

Brandimonte, M.A. and Passolunghi, M.C. (1994). The effect of cue-familiarity, cue-distinctiveness, and retention interval on prospective remembering. *The Quarterly Journal of Experimental Psychology, Section A*, 47(3), 565–87.

Einstein, G.O., McDaniel, M.A., Thomas, R., Mayfield, S., Shank, H., Morrisette, N. and J. Breneiser (2005). Multiple processes in prospective memory retrieval: factors determining monitoring versus spontaneous retrieval. *Journal of Experimental Psychology*, 134(3), 327.

Einstein, G.O., McDaniel, M.A., Williford, C.L., Pagan, J.L. and Dismukes, R. (2003). Forgetting of intentions in demanding situations is rapid. *Journal of Experimental Psychology: Applied*, 9(3), 147–62.

Kliegel, M., Martin, M., McDaniel, M.A. and Einstein, G.O. (2001). Varying the importance of a prospective memory task: differential effects across time- and event-based prospective memory. *Memory*, 9(1), 1–11.

Kliegel, M., Martin, M., McDaniel, M.A. and Einstein, G.O. (2004). Importance effects on performance in event-based prospective memory tasks. *Memory*, 12(5), 553–61.

Logie, R.H., Maylor, E.A., Della Sala, S. and Smith, G. (2004). Working memory in event-and time-based prospective memory tasks: effects of secondary demand and age. *European Journal of Cognitive Psychology*, 16(3), 441–56.

Marsh, R.L., Hicks, J.L. and Cook, G.I. (2005). On the relationship between effort toward an ongoing task and cue detection in event-based prospective memory. *Journal of Experimental Psychology: Learning Memory and Cognition*, 31(1), 68–74.

McDaniel, M.A. and Einstein, G.O. (2000). Strategic and automatic processes in prospective memory retrieval: a multiprocess framework. *Applied Cognitive Psychology*, 14(7), S127–S144.

McDaniel, M.A., Guynn, M.J., Einstein, G.O. and Breneiser, J. (2004). Cue-focused and reflexive-associative processes in prospective memory retrieval. *Journal of Experimental Psychology: Learning Memory and Cognition*, 30(3), 605–14.

Nowinski, J.L. and Dismukes, K. (2005). Effects of ongoing task context and target typicality on prospective memory performance: the importance of associative cueing. *Memory*, 13(6), 649–57.

Oprins, E., Burggraaff, E. and van Weerdenburg, H. (2006). Design of a competence-based assessment system for air traffic control training. *The International Journal of Aviation Psychology*, 16(3), 297–320.

Smith, R.E. (2003). The cost of remembering to remember in event-based prospective memory: investigating the capacity demands of delayed intention performance. *Journal of Experimental Psychology: Learning Memory and Cognition*, 29(3), 347–60.

Smith, R.E. and Bayen, U.J. (2004). A multinomial model of event-based prospective memory. *Learning, Memory*, 30(4), 756–77.

Smith, R.E. and Bayen, U.J. (2005). The effects of working memory resource availability on prospective memory: a formal modeling approach. *Experimental Psychology*, 52(4), 243–56.

Stone, M., Dismukes, K. and Remington, R. (2001). Prospective memory in dynamic environments: effects of load, delay, and phonological rehearsal. *Memory*, 9(3), 165–76.

Chapter 3

Analysis of Learning Curves in On-the-Job Training of Air Traffic Controllers

Esther Oprins
TNO Defense, Safety & Security, The Netherlands

Ernst Burggraaff
Air Traffic Control the Netherlands (LVNL), The Netherlands

Robert A. Roe
Maastricht University, Dep. of Organization & Strategy, The Netherlands

Introduction

This chapter describes a competence-based assessment system, called CBAS, for air traffic control (ATC) simulator and on-the-job training (OJT), developed at Air Traffic Control The Netherlands (LVNL). In contrast with simulator training, learning processes in OJT are difficult to assess, because the learning tasks cannot be planned in advance due to the ongoing air traffic. The assessment system in OJT was designed in such a way that the trainees' progression can nonetheless be monitored. The reliability and validity of CBAS have been evaluated in previous research (Oprins, Burggraaff, and Van Weerdenburg, 2006, 2008; Oprins, 2008). Here we present an evaluation with regard to the analysis of learning curves derived from assessment results. An adequate assessment of learning processes showing differences in individual learning patterns (e.g., slow starters, learning plateaus) and in performance (strengths and weaknesses) is of crucial importance as a basis for feedback on the learning process and for passfail decisions. CBAS compares the trainees' actual performance to the required performance at successive moments of time. Under the assumption that performance is a reflection of the learning process (Oprins, 2008), it extracts learning curves from a sequence of performance measures over time. If trainees are learning, then their performance will increase (Oprins, 2008).

In this chapter we begin by describing the main principles of CBAS and the use of learning curves in OJT. Next, we explain the method by which learning curves are derived from the assessment results. Finally, we present the results of the analysis of learning curves.

The Competence-based Assessment System (CBAS)

The notion of competence refers to the successful integration of knowledge, skills and attitudes, and their application in realistic environments (Oprins, 2008). Competences indicate an individual's ability to effectively perform certain tasks when situational factors are held constant. As competences are based on learning, the assessment of competences at a particular moment in time is essential for adequate feedback and needed to improve individual performance. The first step in the design of the competence-based assessment system (CBAS) at LVNL was a competence analysis based on input from a peer group of air traffic controllers and literature research. The way in which the competence analysis was carried out has been described elsewhere (Oprins, Burggraaff, and Van Weerdenburg, 2006). Here we describe the resulting ATC Performance Model that has served as the framework leading to the design of CBAS (see Figure 3.1).

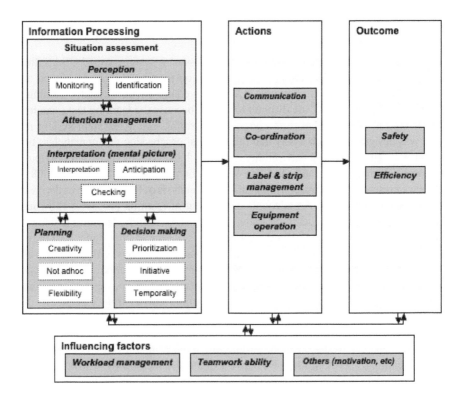

Figure 3.1   ATC performance model

In CBAS, each competence is represented by a set of *performance criteria*. The criteria are formulated in the jargon of the air traffic controllers in order to maximize comprehension and recognition of behaviors. They have been given the format of "behavioral markers" (O'Connor, Hormann, Flin, Lodge, and Goeters, 2002) and are rated on a six point scale. The use of these performance criteria is important as they allow controllers, in their role of assessors, to express their "gut feeling", and to specify why a trainee performs (in)sufficiently and what should be done by way of individual training interventions.

During training the same competences are assessed a number of times, in order to follow trainees' progression over a certain time period and to gain a better insight into deficiencies in performance in various task situations. In ATC training, competences are assessed against augmenting *performance standards*. More specifically, simulator training and OJT are divided into phases, each having its own standards. The standards have been formulated as "behavioral examples", which resemble the "behavior anchors" used in behaviorally anchored scales (BARS; see Berk, 1986). However, they do not specify scale positions but standards to be achieved at the end of each phase. In this way, it is clearer to assessors what is expected from trainees in intermediate phases. For the trainees it is clearer which competences they have to develop further in a specific phase. While the phases in simulator training have mainly been defined by the sequence of simulator exercises, there is no fixed structure in OJT due to the ongoing live traffic. The OJT is divided into four phases following these three main principles: (a) the requirements during training move from *safety* towards *efficiency*, with (b) increasing *traffic complexity*, and with (c) decreasing *aid of the coach* because of acquired expertise. The phases are designated as:

1. Familiarization phase
2. Learning phase 1
3. Learning phase 2
4. Consolidation phase.

The length of each phase is flexible, dependent on trainees' progression. See Table 3.1 for an example of *building a mental picture* in OJT of area control (ACC). Five *performance criteria* (see left column) are rated at a six point scale for assessing the competence, *building a mental picture*. The *performance standards* that belong to each phase, that is, familiarization phase, learning phases 1 and 2, and the consolidation phase, are presented as behavioral examples in the next four columns. The aforementioned three principles can be recognized in the examples.

To get a complete picture of trainees' performance and to measure progression over time, CBAS uses continuous assessment of performance in retrospective progression reports, using the performance criteria of the ATC performance model. The progression reports are completed after one to two weeks of training. Multiple assessors are involved in both types of assessment for maximizing reliability. A web-based assessment tool is used to complete progression reports,

Table 3.1   Example of performance standards for building a mental picture in ACC OJT

Building a mental picture	Familiarisation Phase	Learning Phase 1	Learning Phase 2	Consolidation Phase
Keeps a clear overview of the traffic situation by scanning regularly Looks, observes and takes action if necessary	As the trainee is still getting used to live surroundings he is inclined to focus too much on certain conflicts or flights instead of scanning the entire sector regularly for other traffic requesting attention. Therefore it may happen that a certain flight is not cleared further. He may also be surprised by the call of an aircraft.	During normal traffic the trainee has a good picture of the traffic situation as he scans regularly. He is no longer surprised by calls of any aircraft as he doesn't only focus on specific conflicts or flights, but scans the entire sector regularly. Only during complex traffic may he still be inclined to lose his overview.	During normal and complex traffic situations the trainee has a good overview of the traffic because he is scanning regularly.	The trainee works independently as stated in Learning phase 2 and has acquired sufficient experience at the end of this phase to take the practical exam.
Checks available information to be correct Guards the identification process of the label presentation	At this stage the trainee finds it hard to react at everything he sees and hears. There is more visual and audible information than in the simulator. Therefore it is difficult to act adequately when traffic gets heavier.	By watching changes in information closely, listening to calls by aircraft processing information from the sector(s) and taking action where necessary the trainee is in control during normal traffic situations. He is able to concentrate during a longer period of time and is pro-active.	By watching changes in information closely, listening to calls by aircraft processing information from the sector(s) and taking action where necessary the trainee is also in control during complex traffic situations. He is able to concentrate during a longer period of time and is pro-active.	
Anticipates future and variable traffic situations	The trainee finds it hard to estimate intermediate FLs, the sequence of aircraft (e.g. diverging/converging tracks, speeds, and influence of the wind). At this stage the trainee finds it hard to divide his attention well and have a complete image of the traffic at all times, as he still has to get used to the live surroundings. Therefore it may happen he doesn't discover a flight without a label in time. And when he fits the strip into the sequence he doesn't realise the SSR code has to be adapted. Sometimes the coach has to remind him to change the SSR code or have it changed.	The trainee is monitoring all the time checking that the traffic situation develops as expected, e.g. after giving certain instructions, checking the wind influence and the pilot's reaction to the instructions. Only in complex situations the trainee finds it hard to estimate intermediate FLs, the sequence of aircraft (e.g. diverging/converging tracks, speeds, influence of the wind). During normal traffic situations the trainee can create a good mental picture of all traffic, also of flights that have not been identified by the system. He links the flightplan to the radarposition in time and correctly, the more so as he has remarked on the strips that the SSR code has to be adapted.	The trainee has no problems estimating intermediate FLs, the sequence of aircraft (e.g. diverging/converging tracks, speeds, and influence of the wind). During any traffic situation (also complex) the trainee can create a good mental picture of all traffic. Also of flights that have not been identified by the system. He links the flight plan to the radar position in time and correctly, and sees to it that every flight has the correct SSR code.	
	The coach is closely involved in the handling of traffic and gives tips regularly for a safe and efficient traffic flow. The coach asks many questions like" what will you do with that information, what does this information mean", etc. During heavier traffic the coach can take over control completely for a while.	The coach still gives tips to increase efficiency and only has to intervene in complex situations when safety is at stake.	The coach gives tips only occasionally for efficiency reasons and intervenes only rarely during complex situations and if safety requires.	The coach is only present as the person who is formally responsible for the safety.

to store the results in a database, and to generate overviews of trainee performance – information that is useful for monitoring trainees' progression. Passfail decisions are based on the trainees' progression over time. Trainees go on to the next phase if all performance ratings are rated with a value four or more, which serves as the required standard. Trainees can stay longer in a phase if necessary, but they fail if they do not show any progression over a longer period of time. Passfail decisions are not based solely on a quantitative measure, but they are primarily based on expert judgement of training managers. Continuous assessment with rating the same competences at subsequent moments in time makes it possible to derive learning curves from the assessment results.

Learning Curves in CBAS

Learning Curves and Learning Theory

Learning curves are usually presented as growth curves measuring performance of the same task execution at successive moments in time. Their purpose is modelling learning processes. A simplified ATC task, the Kanfer-Ackerman task, has often been used for examining complex skill acquisition (e.g., Ackerman, 1989; Lee and Anderson, 2001; Taatgen and Lee, 2003). General learning theory says that each learning curve ends in an asymptote (cf. learning plateau) in compliance with the "power law of practice" (Newell and Rosenbloom, 1981). See Figure 3.2.

The growth curve of learning, as shown in Figure 3.2, especially applies to simple skill acquisition. Learning curves can take different forms depending on task complexity, task consistency, and individual differences in learning. In the case of complex tasks, learning curves are different. Complex tasks pose higher demands on cognitive abilities than simple tasks (Ackerman, 1989) and they typically comprise consistent and non-consistent task components (Ackerman, 1989; Schneider, 1990). Consistent task components show large improvements through practice whereas non-consistent tasks do not (Schneider, 1990): consistent

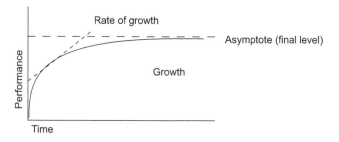

Figure 3.2   Growth curve of learning

tasks require automation while non-consistent tasks require more controlled information processing. A trainee needs time to assimilate new knowledge and skills with previous experiences in order to automate skills. As a result, the learning process of each individual has asymptotes or intermediate learning plateaus. The ATC task can be subdivided into many task components. Under the assumption that these task components obey the "power law of practice", and that they are learned one after each other, we would expect that the overall learning curve would consist of a sequence of smaller learning curves for each task component to be learned in accordance with the findings of Lee and Anderson (2001). This is visualized as the learning curve in Figure 3.3.

However, in on-the-job training skill, acquisition is likely to be more complex. Learning is strongly dependent on the quality of coaches and other influences such as the mental pressure to succeed in training or the complex working environment (physical conditions, colleagues, etc.). In OJT, learning tasks are usually not delivered in a pre-structured sequence. Also, many tasks are trained simultaneously as trainees should be able to manage multiple tasks at the same time. Learning curves would therefore be less volatile than visualized in Figure 3.3, because they can be seen as the result of many overlapping curves. In addition, the order and tempo of learning may differ across trainees because of individual differences in underlying factors (cognitive ability, learning style, personality, pre-education, external influences, etc.). Figure 3.4 illustrates how some variations of learning curves, smoother than presented in Figure 3.3, are expected in complex skill acquisition in OJT with live traffic, under the assumption that all trainees start at the same (zero) performance level.

Figure 3.4 presents a fast learner who achieves the final performance level earlier than on average (dotted, highest curve), a so-called "slow starter" who needs more time and who shows an intermediate learning plateau (continuous, middle curve), and a learner who never achieves the required level and probably fails (striped, lowest curve). Ideally, the assessment system should optimally reflect the kind of learning curves depicted here, but our assessment system cannot produce these learning curves in the same way.

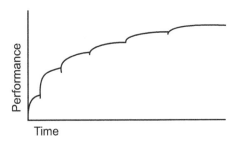

Figure 3.3   Learning curve of complex skills

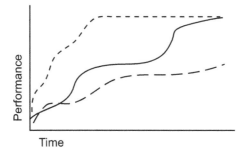

Figure 3.4   Three variations of learning curves

Recalibrated Learning Curves

The learning curves produced by CBAS are based on a weighed sum of competence ratings, however, they are not growth curves as commonly applied. First, the standards against which the trainee is assessed are constantly changing, while remaining mapped on the same six point rating scale. This means that this rating scale, with a value of four or more being sufficient, is constantly being recalibrated. When ratings would stay "sufficient" over time, this implies that the trainee is succeeding in meeting the increasing standards and shows progression. Second, the recorded measurements are not really continuous since they only reflect performance at specific moments. In between these moments, distinct learning processes can take place that cannot be captured completely. In addition, the intervals of measurement differ among trainees due to different training schedules. In this study, the moments in time are presented as rank orders to be able to compare learning curves among trainees. For these reasons, the "learning curve" produced by our assessment system can only be seen as a derivative of the real learning curve. Figure 3.5 visualizes these recalibrated learning curves as produced by CBAS.

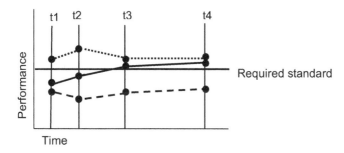

Figure 3.5   Recalibrated learning curves, produced by CBAS

The dotted, highest learning curve represents a trainee who constantly performs above standards, the continuous middle curve reflects a trainee with a temporary learning plateau around t1 and t2 ("slow starter"), and the striped lowest curve refers to a trainee who constantly performs below standard and fails. Straight lines connect the points of measurement to obtain a certain learning curve, but these lines do not necessarily represent the real learning processes in-between. In Figure 3.6 the three trainees are assessed at the same four moments to illustrate possible variations, but in reality trainees are not assessed simultaneously and the intervals differ.

Analysis of Learning Curves

Goal

The analysis of learning curves, as part of the evaluation of the assessment system CBAS, focuses on the representativeness of the learning curves for learning processes. If the assessment results represent learning processes optimally, then CBAS could be applied as an instrument to gain insight into the learning processes of individuals. Adequate feedback could be given, useful interventions could be made (e.g., optimal task selection, coaching and remedial teaching), and passfail decisions could be more valid. Therefore, the main goal of the analysis is to find patterns in the learning curves which are representative for learning.

The assessment system should not only be able to represent general learning processes but also competence development over time. Singular competences are even more important for supporting individual learning: they help to identify deficiencies of trainees required for appropriate feedback and interventions. Trainees may differ in the development of specific competences over time. Some competences may be more trainable than others. Thus, we did not only explore patterns in general performance, but also in the various rated competences.

Design

In order to test whether the recalibrated learning curves differentiate between trainees with varying degrees of training success, we designed a study in which expected "prototypical" learning curves were compared with actual learning curves. Three training managers of LVNL made an a priori classification of trainees in *high performers* (passed without problems), *moderate performers* (passed with difficulties), and *low performers* (failed), based on expert judgment. For each class of learning success we defined prototypical learning curves which serve as hypotheses in the analyses. Figure 3.6 presents these prototypical learning curves in terms of expected zones for the three types of trainees.

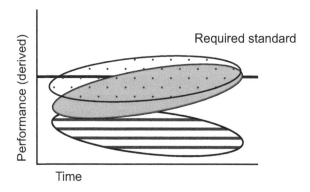

Figure 3.6   Expected zones for learning curves of high (dotted), moderate (grey) and low (striped) performers

The entrance level of trainees varies because of individual differences. Trainees in the low performance category are likely to start at a lower performance level because they also performed lower in the preceding simulator training. The high and moderate performers, who both pass training, are expected to achieve the final standards in the end in contrast with the low performers or failures. This latter group is expected to show decreasing performance compared to the increasing standards, the cumulative nature of the learning process, and a lack of self-confidence as a result of their low performance. Because trainees are assessed against phase level and transfer to the next phase when all competences are rated as sufficient, this scenario is expected in each learning phase of OJT.

Method

We used 403 progression reports made for 27 trainees in OJT of area control (ACC). We did the same analyses only for learning phases 1 and 2 because in these longest phases trainees learn most. We visually examined the learning curves of trainees, classified into the three groups, to investigate patterns in learning processes. We did a discriminant analysis to check whether the classification into the three groups was correctly predicted in a quantitative way. Therefore, we defined four variables, divided into two groups, because we did not know yet which measure would be most suitable for defining learning curves quantitatively:

- *Performance level: 1) mean performance level* (weighted sum of competence ratings); and *2)* occurrences of insufficient performance (value of this weighted sum < 4).
- *Progression: 1) growth* (final minus initial performance level); and *2) rate of growth* (beta coefficient of the linear regression model, obtained by curve fitting).

An analysis of variance (ANOVA) between the three groups, separately for learning phase 1 and 2, was mentioned to provide insight into the differences in means of the variables across the groups.

In addition, we examined to which extent the three groups differ in means on each competence, both phases together, visually and with an analysis of variance (ANOVA). We also calculated rank order correlations between the variable "time" and the competence ratings for each group to investigate the differences in progression, separately for two phases in OJT.

Results

Learning Curves

The derived learning curves for the three groups resemble the prototypical learning curves as presented in Figure 3.6 to a certain extent. See Figures 3.7 and 3.8 for respectively learning phase 1 and 2.

Low performers learning phase 1:

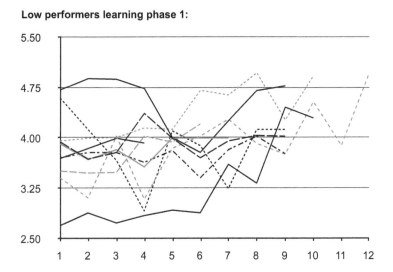

Figure 3.7 **Derived learning curves for respectively high, moderate and low performers in learning phase 1**

Moderate performers learning phase 1:

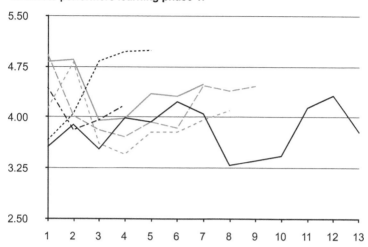

High performers learning phase 1:

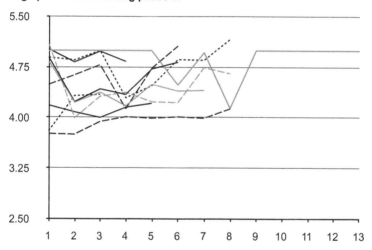

Low performers learning phase 2

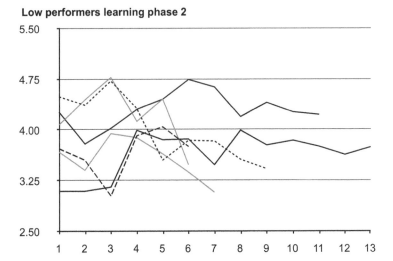

Figure 3.8   Derived learning curves for respectively high, moderate and low performers in learning phase 2

Moderate performers learning phase 2

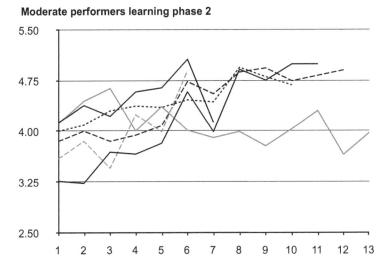

High performers learning phase 2

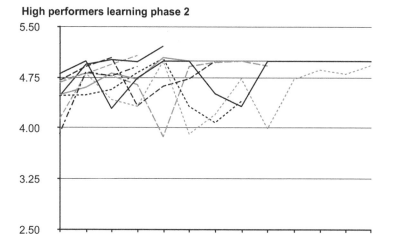

The graphs for the three groups are quite similar for learning phases 1 and 2, as expected, because the same assessment and training methods are used. The group of high performers (N = 10) performs rather constantly above the sufficient standards. We would expect an increase towards the required standards, but this is only visible for moderate and low performers. Apparently, some competence ratings can be insufficient in the beginning, but the weighted sum of ratings stays sufficient for high performers. As expected, the variation between trainees within the group of moderate performers (N = 6) is really high. Trainees' performances approximate the standards but with many "highs and lows". It should be noticed that the low performers (N = 11) are the same trainees as in learning phase 1 and 2; some trainees fail already in phase 1 but others transfer to phase 2 and fail eventually. This explains why the results are more positive for phase 1 and why a smaller number of learning curves is presented for phase 2. However, all graphs present some exceptions which do not fit the patterns well, and the lines are not continuous. This is probably influenced by the unreliability of assessors' ratings.

Discriminant Analysis

For learning phase 1, the first discriminant function is significantly different across groups ($X^2 = 22.36$, df = 8, p = .004). The discriminant function coefficients indicate that *insufficient performance* is the best predictor for the classification into groups, respectively followed by *mean performance level*, *rate of growth* and *growth*. In total, the discriminant function successfully predicted group membership for 73.1 percent, see Table 3.2.

**Table 3.2	Predicted group membership for learning phase 1 (n = 26)**

	Low performers	Moderate performers	High performers	Total
Low performers	7	1	2	10
Moderate performers	1	4	1	6
High performers	1	1	8	10

For learning phase 2, the first discriminant function is also significantly different across groups (x^2 = 45.49, df = 8, p = .000). The discriminant function coefficients indicate that *mean performance* is the best predictor for the classification into groups, respectively followed by *insufficient performance, growth* and *rate of growth*. In total, the discriminant function successfully predicted group membership for 90.5 percent, see Table 3.3.

**Table 3.3	Predicted group membership for learning phase 2 (n = 21)**

	Low performers	Moderate performers	High performers	Total
Low performers	5	1	0	6
Moderate performers	0	6	0	6
High performers	0	1	8	9

Analysis of Variance

An analysis of variance (ANOVA) shows that the means of two variables differ significantly (p < .05) across the three groups for learning phase 1 as presented in Table 3.4; Levene's test showed that the variances of the four variables are all homogeneous.

**Table 3.4	Analysis of variance (ANOVA) for OJT learning phase 1 (n = 26)**

	F	df1	df2	Sig.
Mean performance	8.154	2	23	.002
Insufficient performance	.124	2	23	.884
Growth	1.487	2	23	.247
Rate of growth	14.456	2	23	.000

Rather comparable results were found with an analysis of variance (ANOVA) for learning phase 2. Table 3.5 shows that the means of all variables differ significantly (p < .05) among the three groups. Levene's test showed that we may not assume that the variances of *mean performance level* and *insufficient performance* are homogeneous at p < .05.

Table 3.5   Analysis of variance (ANOVA) for OJT learning phase 2 (n = 21)

	F	df1	df2	Sig.
Mean performance	28.566	2	18	.000
Insufficient performance	3.812	2	18	.042
Growth	7.640	2	18	.004
Rate of growth	8.960	2	18	.002

The results for both learning phases suggest that *mean performance level* and the *rate of growth* are the best predictors for the classification into the groups, although *growth* and *insufficient performance* generally also contribute to this classification. It should be noted that the classification into groups is the same for both learning phases. This may have affected the results because some trainees may fit better in two different groups for the two phases.

Differentiation in Competence Ratings

Next, we investigated to which extent the three groups differ in means on each competence for the two learning phases together. Figure 3.9 shows that competences that seem to be critical for ATC performance, *safety*, *efficiency*, *mental picture*, *attention management*, *planning* and *workload management* are very distinctive. These results are confirmed by an analysis of variance (ANOVA): the means of these competences differ significantly (p < .001) among the three groups in contrast with competences that seem to be less ATC-related such as *equipment operation*, *attitude* and *team orientation.*

Rank Order Correlations

Finally, we examined competence development for each group of performers by calculating rank order correlations between the variable "time" and competence ratings. Table 3.6 generally shows positive correlations. This can be explained by the fact that trainees are assessed against phase level. The low performers who fail in learning phase 2 show progression in learning phase 1 towards the required level, otherwise they would have failed in learning phase 1. The fact that the requirements in learning phase 2 are higher may explain why moderate

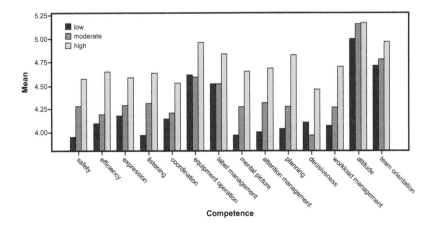

Figure 3.9   Mean competence ratings for two learning phases (n = 27)

performers in learning phase 2 show more progression than in learning phase 1. Thus, these findings confirm the results of the learning curve analyses. Differences between competences are not clear; only label management and team orientation do not show much variation.

Table 3.6   Rank order correlations (Spearman) between time and competence ratings for learning phases 1 and 2

Competence	High performers		Moderate performers		Low performers	
	Phase 1 (N = 71)	Phase 2 (N = 81)	Phase 1 (N = 46)	Phase 2 (N = 64)	Phase 1 (N = 85)	Phase 2 (N = 46)
Safety	.12	.23*	−.25	.37**	.21*	−.22
Efficiency	.33**	.28**	−.09	.50**	.27*	.00
Verbal expression	.24*	−.07	−.00	.42**	.18*	.11
Listening	.23*	.15	−.16	.25*	.18*	.13
Co-ordination	.19	.28**	−.20	.27*	.17	.10
Equipment operation	.05	.03	−.06	.30**	.28**	.01
Label management	.24*	.13	.06	.20	.26**	.05
Mental picture	.22*	.26*	.11	.41**	.31**	.02
Attention management	.31**	.39**	.01	.45**	.26**	−.07
Planning	.13	.08	−.11	.30**	.31**	.04
Decisiveness	.12	.38**	−.10	.48**	.22*	.11
Workload management	.28*	.04	.22	.51**	.10	−.04
Attitude	−.16	−31**	−.06	.33**	.18	.08
Team orientation	.03	.07	.03	.08	.10	.16

** Correlation is significant at the 0.01 level (1-tailed); * Correlation is significant at the 0.05 level (1-tailed)

Discussion and Conclusions

Representative Learning Curves

This chapter proposes a new method to derive learning curves from assessments. We made "recalibrated" learning curves based on assessment against performance standards that are constantly recalibrated into the same rating scale. This was needed to examine whether learning curves, produced by the assessment system, were sufficiently representative for learning processes as part of the evaluation. In this way, this study adds to the studies with the Kanfer-Ackerman task, also an ATC task, aimed at modelling learning processes (Ackerman, 1989; Lee and Anderson, 2001; Taatgen and Lee, 2003).

The results have shown that the assessment system is able to represent patterns in learning processes sufficiently. This provides evidence of the quality of the assessment design. We distinguished three groups of trainees based on training success. Their recalibrated learning curves derived from assessment results agree with the three defined patterns of prototypical learning curves in conformance with general learning theory (Newell and Rosenbloom, 1981). Performance usually improves with practice and therefore high and moderate performers show progression over time in the recalibrated learning curves. Low performers achieve a learning plateau earlier than the moderate and high performers. However, the graphs show many highs and lows and they present a certain number of exceptions. This confirms that ATC is a very complex skill to acquire (Schneider, 1990), especially in live environments. In this context, we should also take into account that the assessors' ratings are not completely reliable (Oprins, Van Weerdenburg, and Burggraaff, 2008; Oprins, 2008). This is very difficult to achieve in OJT, more than in simulator training, due to the ongoing variety of task situations, multiple assessors and many individual differences in learning.

The quantitative analyses with four variables (discriminant analysis, analysis of variance) have shown that classification into the three groups was correctly predicted. The best predictors were *mean performance level* (weighted sum of competence ratings) and *rate of growth* (beta coefficient of the linear regression model). *Mean performance level* is probably a better measure than *insufficient performance* because it indicates how much the trainee performs below or above the standards instead of how often. For progression, *growth* (difference between final and initial performance level) is less distinctive than *rate of growth* because individual learning curves show many variations caused by unstable performance of trainees. Rate of growth indicates a certain direction of the learning curve based on the whole range of measures. However, it should be noticed that the beta coefficients are not always reliable since the linear regression model does not fit for each trainee. We also should realize that progression in the recalibrated learning curves differs from that in general learning curves: a horizontal line implies that the trainee is still learning since the required standards increase over time.

Competence Development

The three groups of trainees differ much in mean competence ratings. The largest differences were found in competences that are assumed more critical and less trainable such as *mental picture* (cf. Situational Awareness) and *workload management* in comparison with less critical and more trainable competences such as *equipment operation* and *attitude*. This relates to the distinction between respectively "consistent task components" that depend heavily on individual abilities and "non-consistent task components" that improve by more practice (Schneider, 1990). ATC is a combination of both. Our findings suggest that some competences are more trainable than others. The three groups also differ in progression on competences. The findings support the analyses of learning curves in a quantitative way.

Practical Implications

These findings have some practical implications. The learning curves can be used to adapt training to the trainees' needs, for instance, by recognizing slow starters and (intermediate) learning plateaus. Following progression on singular competences can help to detect specific deficiencies of trainees and to repair them as a next step. More insight should be gained in how to make training adaptive, for instance, by (dynamic) task selection, specific coaching, remedial teaching, re-training, etc. Development of self-directed learning skills might help trainees to define what they need by themselves. In addition, passfail decisions can be improved by using learning indicators with sufficient predictive validity. Trainees can be classified in one of the three groups based on the characteristics of their (recalibrated) learning curve, and, finally, predictions can be made about future learning. Ultimately, the measures *mean performance level* and *rate of growth* can be used as a cut-off for passfail if predictive validity will have been proven.

General Conclusion

In sum, patterns in learning processes can be clearly recognized in the recalibrated learning curves produced by the assessment system CBAS applied in OJT. This implies that CBAS is a well-designed instrument to follow trainees' progression over time, to provide adequate feedback and to develop effective training interventions. The next step is to use the learning curves for improving pass/fail decisions based on quantitative performance measurement. Therefore, research on learning curves and how to derive them from assessment results must be continued, involving a higher number of trainees.

References

Ackerman, P.L. (1989). Individual differences and skill acquisition. In: P.L. Ackerman, R.J. Sternberg and R. Glaser (eds), *Learning and Individual Differences: Advances in Theory and Research*, (pp. 165–217). New York: Freeman and Company.

Berk, R.A. (1986). *Performance Assessment: Methods and Applications.* Baltimore: The Johns Hopkins University.

Lee, F. and Anderson, J.R. (2001). Does learning a complex task have to be complex? A study in learning composition. *Cognitive Psychology*, 42, 267–316.

Newell, A. and Rosenbloom, P.S. (1981). Mechanisms of skill acquisition and the law of practice. In: J.R. Anderson (ed.), *Cognitive Skills and Their Acquisition*, (pp. 1–55). Hillsdale, NJ: Erlbaum.

O'Connor, P., Hormann, H., Flin, R., Lodge, M. and Goeters, K. (2002). Developing a method for evaluating crew resource management skills: a European perspective. *The International Journal of Aviation Psychology*, 12(3), 263–85.

Oprins, E. (2008). Design of a competence-based assessment system for air traffic control training. PhD thesis, Maastricht University.

Oprins, E., Burggraaff, E. and Van Weerdenburg, H. (2006). Design of a competence-based assessment system for air traffic control training. *The International Journal of Aviation Psychology*, 16(3), 297–320.

Oprins, E., Burggraaff, E. and Van Weerdenburg, H. (2008). Reliability of assessors' ratings in competence-based air traffic control training. *Human Factors and Aerospace Safety*, 6(4), 305–22.

Schneider, W. (1990). Training high-performance skills: fallacies and guidelines. In: M. Venturino (ed.), *Selected Readings in Human Factors*, (pp. 297–311). Santa Monica, CA: The Human Factors Society.

Taatgen, N.A. and Lee, F.L. (2003). Production compilation: a simple mechanism to model complex skill acquisition. *Human Factors*, 45(1), 61–76.

Chapter 4

How Cockpit Crews Successfully Cope with High Task Demands

Ruth Haeusler
Zurich University of Applied Sciences, Switzerland

Ernst Hermann
University of Basle, Switzerland

Nadine Bienefeld
Federal Institute of Technology, Switzerland

Norbert Semmer
University of Berne, Switzerland

Introduction

Choosing an adequate strategy saves limited resources such as time and cognitive capacity when striving for a goal. Performance in the cockpit depends strongly on contextual aspects such as specific tasks (e.g., precision or non-precision approach), crew composition (e.g., level of experience), operational environment (weather conditions, airport layout, etc.), interaction with other groups such as air traffic control, and dynamic factors such as potential technical failures. The significance of contextual factors for performance makes adaptation to the specific operational demands a key factor for performance. Research on cockpit crews confirms that adaptation to contextual factors is significant for performance: Crew Resource Management (CRM performance) of cockpit crews varies across different scenarios, indicating low trans-situational consistency (Haeusler, Klampfer, Amacher and Naef, 2004; Brannick, Prince, Prince and Salas, 1995). In other words: one cannot assume that crews trained in one type of scenario will automatically transfer the acquired skills to another scenario.

On the other hand, complex and dynamic work environments such as the cockpit require people to flexibly adjust their task performance strategies to the particular demands of the situation. Therefore, identifying the relevant aspects of a given situation and adapting to these situations in a flexible way is important (Pulakos, Arad, Donovan and Plamondon, 2000). Research on expert performance indicates that consistently excellent performance is based on the availability of distinctive task performance strategies and a highly structured knowledge of the relevant domain. These prerequisites ("routine expertise"; Holyoak, 1991; Smith, Ford, and Kozlowski, 1997) enable experts to develop "adaptive expertise", that is, to adapt their actions to the various task demands, to anticipate adequately future developments and to react quickly when the situation changes. The degree to which cockpit crews adapt to abnormal situations for instance, is predicted by such crew behavior as "search and transfer of information", "prioritizing of tasks" and "distribution of tasks" (Waller, 1999). Such adaptation implies explicit planning; furthermore, adaptation is likely to profit from meta-cognitive activities such as encoding situational characteristics and developments into more general rules (cf. Holyoak, 1991).

The first part of this chapter integrates theoretical concepts and empirical findings from Human Factors and work psychology (most notably action theory and evolution of expertise) in particular to explain the capability to successfully cope with high task load. The second part presents an overview of the results of a field study that was conducted to test hypotheses about how cockpit crews deal with complex and novel tasks successfully (Haeusler, 2006). Forty-five cockpit crews were observed in respect to their adaptation to situational demands in two scenarios with moderate and high task loads. The result was that processes of extensive orientation regarding situational demands (situation awareness) and thorough planning correlated significantly with behavior that was optimally adapted to the situation (high degree of adaptation). However, the advantage of these preparatory processes was confirmed only in one of the two scenarios when task load was moderate. Meta-cognitive activity further enhanced the degree of adaptation, but again only in the scenario with moderate task load. In the scenario with high task load a more reactive pattern of adaptation prevailed: monitoring and coaching were the best predictors for the degree of adaptation. Benefits of the selection of adaptive task strategies, which specifically facilitate coping with situational demands, were found for technical performance in both scenarios. As expected, the general strategy of augmenting effort (trying harder) in the light of challenging tasks did not enhance technical performance for most of the indicators, with the exception of controlling aircraft speed. Some moderator effects of adaptation were also found: increased effort led to higher technical performance only when combined with adaptive task strategies. Contrary to our expectations subjective workload of the crews with a high degree of adaptation was not lower. Results indicate that mental workload is critical: exhaustion of capacity by more complex tasks and problems may critically restrict the ability to proactively adapt to situational demands. A more reactive pattern of adaptation through monitoring and coaching is predominant. Adaptation itself is a process demanding cognitive resources that are highly charged with complex tasks. The relevance of these results and practical implications to stimulate adaptation in training are discussed at the end of this chapter.

Performance Strategies Under High Task Load

Research on performance under high task load has repeatedly failed to indicate clear signs of performance decrements. While this may partly be due to methodological problems in measuring performance (cf. Hockey, 2002), findings show that people actively adjust to changing demands (e.g., Sperandio, 1978; Sauer, Hockey and Wastell, 1999). They prioritize and/or simplify the handling of tasks to cope with challenging situations. Being able to reduce demands on limited resources by adapting one's performance strategy is essential to cope with high task load. Strategies are defined as a deliberate choice of the course of action to reach a goal. Since the capacity to process information and simultaneously execute tasks is limited, strategies attempt to track a target with the means and resources available at the moment. They represent an act of self-regulation to modulate the workload. Strategies can be more or less useful in meeting the goal (effectiveness) with many or few resources (efficiency). Strategies determine the choice, prioritization and scheduling of tasks (Segal and Wickens, 1991). Scheduling strategies strive to manage resources to cope with high task load by prioritizing and scheduling tasks. This allows operators to keep the workload at an acceptable level. At the same time, attention and effort are focused on primary tasks, which protects task performance at the expense of less important tasks (Hart, 1989; Hart and Wickens, 1990; Hockey, 1997). Cockpit task management (Funk, 1991) represents a prominent approach to investigate the scheduling strategies of pilots. It involves initiating tasks, allocating resources to tasks, monitoring task execution, prioritizing, interrupting, continuing and then terminating tasks. This chapter specifically looks into the question of how these strategies can aid in adapting to situational demands.

Control Mechanisms to Adapt the Level of Workload

The variable state activation theory identifies two control mechanisms that are connected to different strategies for coping with task demands (Maule and Hockey, 1993). Routine control mechanisms automatically cope with familiar tasks. A conscious mode of control (supervisory control) is activated by unfamiliar task demands and leads to either (1) increase in effort (trying harder), (2) re-appraisal of goals to reduce demands, (3) altering or eliminating demand (e.g., by negotiating new time limits) and/or (4) doing nothing. In dynamic work environments such as cockpits, control strategies that cope with high task load are of interest. According to Hockey (1997), a compensatory style of regulation may involve: (1) reducing goals for non-critical aspects, (2) optimizing resource use by redirecting attention resources from secondary to primary tasks, and/or (3) increasing effort to enhance primary and secondary task performance. These strategies protect performance in primary tasks at the cost of latent performance degradation. These can occur as (a) decrements in secondary task performance, (b) strategic adjustments to reduce effort in task management, (c) increased activation and task related effort leading to increased costs in task accomplishment, and (d) after-effects such as the selection of less energetic performance strategies in subsequent work periods to regain resources. The strategy of augmenting effort (c) is of limited use, as people get exhausted over time. A study on driving instructors demonstrates performance decrements of the compensatory effort strategy (trying harder) when people get fatigued (Meijman, Mulder, van Dormolen and Cremer, 1992).

The goal is therefore to work smarter, not harder (cf. Sujan, 1986). People make use of the relevant information about the task/problems they are confronted with by adapting their strategy (e.g., Segal and Wickens, 1991; Moray, Dessouky, Kijowski and Adapathya, 1991). Sometimes, however, their workload is higher due to a preoccupation with the task at hand (Moray et al., 1991). Thus, optimal task management goes along with additional cognitive costs. The question therefore is: how can adaptive task strategies be developed while keeping the workload at an acceptable level despite the extra costs of mental preparation?

Adaptation to Cope with Situational Demands

Situational demands arise from task difficulty, the necessity to coordinate with others, contextual factors such as the operational environment (e.g., weather, terrain), and the necessity to deal with sudden problems (e.g., technical failures). For instance, one of our scenarios was characterized by a non-precision approach that was rarely flown (task difficulty) in an environment restricting the choice of options (terrain around the airport) and requiring good coordination within the crew, because the pilot flying could not see the runway during visual circling and therefore had to rely on information by the pilot non-flying (coordination). Adapting to high task load situations focuses on reducing the need for limited mental resources for task execution. By optimally considering situational demands, overload and extra effort such as error management can be avoided. For instance, choosing a higher circling altitude than the minimum descent altitude relieves the pilot flying from permanently monitoring the altitude.

Adaptation to situational demands may be based on a deliberate choice of strategy, on routine actions based on experience, on standard operating procedures, or on chance. Routine actions offer economic ways of fulfilling a task and require few attention resources under consistent task conditions. But in novel situations they can lead to errors and performance decrements, when the particular demands of the task situation are not met. Strategies aim at choosing or modifying the process of getting a task done or a problem solved, with a view to avoid shortages in resources during task execution, while complying with time limitations. Adaptability is characterized by flexibility in selecting a strategy that considers changes in demands. This is in contrast to stable task contexts, where routines allow more automatic processing of information and less control in the regulation of actions.

Concepts of adaptation focus on various aspects: (1) personal characteristics (e.g., personal initiative, Fay and Frese, 2001; proactive personality, Bateman and Crant, 1993) are considered essential for coping with changes of the organizational context (e.g., evolution of technology, global competition, etc.). (2) Characteristic behavior of individuals (e.g., adaptability, LePine, Colquitt and Erez, 2000; Pulakos, Arad, Donovan and Plamondon, 2000) and teams (team adaptability, Kozlowsky, Gully, Nason and Smith, 1999) are described that enhance adaptive task performance strategies in the context of training. Adaptable teams, for instance, successfully avoid overload of central roles within their team and manage to maintain team coordination when task load increases. (3) Task Adaptive Behavior (TAB) is an approach used to measure the congruence of task performance strategies with specific demands inherent to the task and situation (Tschan, Semmer, Naegele and Gurtner, 2000). Factors enhancing task adaptive behavior are described in more detail in the next sections.

Factors Stimulating Adaptation

Five aspects potentially stimulate adaptation: organizational norms and procedures, task characteristics, experience and routines, deliberate initiation and control of actions, and meta-cognitive activity.

Organizational Norms and Procedures

Organizational norms and rules specify a standardized procedure in the light of specific conditions. They are designed to prevent errors from (re)occurring and create transparency and predictability of actions within the organization (Reason, 1990). They can regulate the proceeding or the final state to be reached (Grote et al., 2004). In aviation, standard operating procedures (SOPs) and emergency procedures allow cockpit crews to coordinate work even if the pilots have not flown together before. Sometimes, however, pilots need to adapt a procedure, if it is not adequate for a given situation.

Task Characteristics

The consistency of a task influences performance directly (Maynard and Hakel, 1997) and indirectly by the choice of strategy or by the augmentation of effort (Campbell, 1991; Locke, Shaw, Saari and Latham, 1981). Action control theory defines tasks as goals that need to be achieved by executing certain operations. These operations are limited by prevailing conditions (Frese and Zapf, 1994). How a goal is achieved may differ in terms of effort spent, time needed, and reliability. The results a person can achieve in a certain task depend on his/her skills. Task complexity is relevant for the question of how to handle high task load situations. Complexity is described in objective and subjective terms. (For an overview, see

Campbell, 1988.) Objective task complexity derives from the required actions, the informational cues to be processed, and the demands in respect to knowledge, skills and resources for task accomplishment (Wood, 1986). Tasks can be complex in regard to content, coordination of subtasks, and dynamic changes on the level of content or coordination. Coping with complex tasks requires effortful processes. These correspond with knowledge-based regulation of tasks (Rasmussen, 1986), and draw on limited capacities in regard to attention and working memory. Task demands correspond with task difficulty and task load. They are thought to be high when tasks are complex, dynamic, novel or difficult to perform (e.g., when much and ambiguous information has to be processed and coordination of multiple tasks is needed [multi-tasking]).

Experience and Routine

Experience and routine allow operators to perform tasks with less effortful control (rule- and skill-based; Rasmussen, 1986). This saves attention resources and capacity of the working memory, making actions more resistant to stress (Driskell and Johnston, 1998). This is the basis of routine expertise (Holyoak, 1991), which allows people to reinvest free capacity in deducing and modifying strategies. On the other hand, the availability of automated routines makes people vulnerable to stereotyping errors, because they miss changes and therefore do not adapt to them. Whether routines enhance performance more than they do harm depends on the frequency of changes relevant to performance, the size of the changes and the amount of adaptation needed (Cañas, Quesada, Adoracion and Inmaculada, 2003). Skill- and rule-based actions need little conscious processing of information. Rather than on a comprehensive analysis, the perception of the task and the situation is based on pattern recognition: familiar patterns are activated in long-term memory and the situation is directly associated with solutions stored in memory (cf. recognition primed decision making; Klein, 1997), requiring little or no planning. So the question is: how can people be made aware of novel or changing task demands when they actually tend to recognize primarily only those aspects with which they are familiar (selective attention) and do not feel the need to process more information? This leads to the last two factors that enhance adaptation.

Deliberate Initiation and Control of Actions

Recognizing situational demands is a prerequisite for the selection of adequate strategies. Deliberate initiation and control of actions are important aspects for adaptation since goals need to be redefined in view of novel tasks and/or changes in the task context. A high level of situation awareness is needed to ensure that information about the specific task demands is consciously processed and considered in planning (proactive adaptation; cf. threat management, Helmreich, Klinect and Wilhelm, 1999). The implementation of the plan is further supported by monitoring and coaching the execution of strategies. On the other hand, monitoring and coaching may also compensate for a lack of adaptation (reactive adaptation; cf. error management, Helmreich et al., 1999).

Goal setting and planning are processes of self-regulation. Both require the ability to anticipate implications and future trends and to analyze causal relationships. They focus attention on processing more information and feedback (Sonnentag, 2000). Under high task load planning tends to be reduced or even halted due to the demands on attention resources (Kanfer and Ackermann, 1989). Therefore, it is important to use phases with low task load for planning (Orasanu and Salas, 1993). In addition, detailed planning in the sense of "when situation x, then y" will automate control of actions (Bargh and Gollwitzer, 1994). This enables the implementation of planned actions despite increased task load, because characteristics of the external situation will act as triggers for action. Contingency planning can improve performance (Tripoli, 1998) and decision making (Pepitone, King and Murphy, 1988).

Monitoring and controlling activities enhance goal attainment by focusing attention on target tracking (Kanfer and Ackerman, 1989). Feedback processing compares actual and target levels and allows recognition of errors, inconsistencies and suboptimal strategies. More experienced persons tend to look for more feedback and recheck their solutions more often than inexperienced persons (Sonnentag, 2000). Highly interactive teams use team monitoring, not only to monitor and detect errors, but also to regulate individual inputs in coordinated teamwork (Marks and Panzer, 2004).

Meta-cognitive Activity

Meta-cognitive activity is the awareness and control of one's thoughts (Flavell, 1979). Meta-cognitions activate self-monitoring during task execution, self-evaluation against the target level and self-regulation to achieve a goal (Kanfer and Ackerman, 1989). Meta-cognitive and self-regulatory skills are considered important prerequisites for adaptability (Kozlowski, 1998). Research on adaptive expertise – the capability to transfer one's knowledge and skills to novel situations – suggests that meta-cognitive activity enhances exploration of tasks and situations. Adaptive experts understand the principles of why a strategy works. They can explain under which conditions a particular strategy is likely to work or not. This enables them to better adapt their skills to the restrictions of the task environment. Meta-cognition operates on the basis of knowledge about the task (problem-domain knowledge), about the strategy (procedural knowledge) and about oneself (Berardi-Coletta, Buyer, Dominowski and Rellinger, 1995). It enhances adaptation through monitoring one's performance and learning process, and stimulates transfer of procedural knowledge on novel tasks and problems. Meta-cognition stimulates exploration and deliberate analysis of the situation (situation awareness) and goals (planning) and can enhance the development of strategies (Hacker and Skell, 1993), ensuring that the selected procedure satisfies situational demands (Ford, Smith, Weissbein, Gully and Salas, 1998). A study on medical doctors confronted with an unclear diagnosis finds that explicitly explaining one's statements (e.g., with sentences such as "because ...") result in more accurate diagnosis than when doctors just make statements (e.g., "this must be this ...") (Tschan et al., 2009). Meta-cognitive skills are also important in situations with high task load and stress, because they avoid bias in decision-making and reduce errors (cf. recognition/meta-cognition model, Cohen, Freeman and Wolf, 1996).

Field Study

Participants

The sample consists of 90 airline pilots of the A320 fleet of a medium-sized airline. This corresponds to a participation rate of 60.44 percent. Participation was voluntary. Flight experience was between 5,000 and 10,000 hours for the majority of the captains (71.7 percent) and between 1,000 and 5,000 hours for most of the first officers (45.7 percent). (See Table 4.1.)

Table 4.1   Total flying hours and hours on type

	Captains		First officers	
	Total	**On aircraft type**	**Total**	**On aircraft type**
Up to 500 hours		26.1%	15.2%	39.1
Up to 1'000 hours		17.4%	23.9%	32.6%
Up to 5'000 hours	6.5%	56.5%	45.7%	26.1%
Up to 10'000 hours	71.7%		15.2%	
> 10'000 hours	21.8%			
Missing data				2.2%

Material

Analyses comprise two simulator scenarios from the annual training conducted in 2000. Scenario 1 represents a complex non-precision approach (NDB) with subsequent circling. This kind of approach is rarely flown and can be particularly demanding for inexperienced pilots. Pilots have to monitor the aircraft's lateral and horizontal position and direction with navigation aids and have to use manual thrust. The scenario requires good planning and a timely execution of subtasks. Good use of automation facilitates the task. The first officer is the pilot flying in this scenario. The mean duration of scenario 1 is 17 minutes (SD = 4 min 40 sec.). Task load in scenario 1 is moderate.

In scenario 2, multiple technical failures (loss of flight control computers) occur, leading to severe control problems. The crew has to control the aircraft using only rudder, pitch trim and power. An emergency landing with ILS approach is flown in manual back-up law, which is a non-certified emergency procedure. The problem itself is well defined and the required actions are obvious to the crew. The workload is especially high in regard to psychomotor flying skills. The captain is the pilot flying. The mean duration of scenario 2 is 33 minutes (SD = 6 min 14 sec.). Task load in scenario 2 is high.

Variables

The next section briefly describes the variables; for more details see Haeusler (2006). Observations with the category system were event-based and done by two raters based on video material including flight parameters. Interrater reliability testing on 10 percent of the observation material was sufficiently high (Cohen's Kappa: .81 and higher). Figure 4.1 gives an overview on data collection and analysis.

Adaptation   Based on a hierarchical task analysis (Shepherd, 2001), two subject matter experts (both captains and experienced flight instructors) identified situational demands to be considered in each of the two scenarios and specific adaptive task strategies. This provided the basis for two measurements of adaptation: the degree of adaptation and adaptive task strategies. The *degree of adaptation* to situational demands includes all considerations and precautionary measures to deal with the situational demands of the scenario. Just as an example: briefing a modified go around procedure in scenario 2 because of reduced controllability of the aircraft due to technical malfunctions. On the other hand, *adaptive task strategies* are specific reactions to situational demands with the goal (a) to reduce requirements

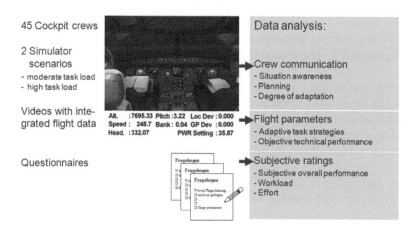

Figure 4.1   Data acquisition diagram

for attention resources (e.g., select higher circling altitude in scenario 1), (b) to manage the workload by delegating tasks to automation (e.g., use autopilot to maintain altitude during circling in scenario 1), by reassigning tasks within the crew (e.g., delegating regulation of thrust to the pilot non-flying in scenario 2) and/ or enlarging team resources (e.g., delegating navigation tasks to ATC by requiring vectoring in scenario 2), and (c) to gain time (e.g., by configuring the aircraft for approach at an early stage). Altogether adaptive task strategies reduce task difficulty. Adaptive task strategies were assessed by flight parameters.

Performance   Overall and specific performance were assessed. To this end, the pilot flying for each scenario had to rate his own performance following the session (*subjective overall performance*). Indicators of technical performance were measured with flight parameters (*objective technical performance*). Technical performance was assessed in scenario 1 with flight parameters for success of approach and falling below critical limits (descending below minimum descent altitude during circling). In scenario 2, flight parameters were analyzed during final approach to measure deviation from target levels in a defined flight segment (deviation from glide path and localizer in the final approach) and variability in performance (standard deviation of pitch, speed and sink rate).

Situation awareness and planning   These cognitive processes were measured based on cockpit crew communication. Situation awareness included remarks that indicated recognition and anticipation of information about the specific task/ problem (e.g., low circling minimum in scenario 1) and external environment (e.g., properties of the runway in scenario 2 in view of an emergency landing) for a comprehensive understanding of the situational demands. Planning includes briefing and preparation of the precise sequence of actions to cope with situational demands, accomplishing the task and handling the problem and possible critical events. Instances of planning are task scheduling (e.g., start of descent at an earlier stage in scenario 2), role clarification (e.g., pilot non-flying intensifies speed control during approach in manual thrust in scenario 1), and contingency management (e.g., preparation for emergency landing in scenario 2).

Monitoring/coaching   This category includes behavioral observations of the pilot non-flying about preparation of actions ("next step is outer marker check"), monitoring of task execution ("you are 100 ft too high"), situation awareness about flight parameters ("your speed is now 160 kts") and error management/correction ("you are too fast; reduce the speed").

Effort and subjective workload   Effort consists of an unspecific mobilization of energy. It can either enhance adaptation or perpetuate inadequate task strategies. Workload expresses the consumption of physical and mental resources and the experience of time pressure while performing a task. Both measurements were taken from the NASA Taskload Index (Hart and Staveland, 1988).

Meta-cognitive processes These were coded in crew communication when hypotheses were generated ("I should be able to tune in to an ILS frequency using the dial on the VOR unit to get a reference for the inbound course 003"), when changes were observed ("I have switched to the backup system and have now received an artificial horizon"), when rules were deducted ("when I put down the nose [in manual backup law] then speed increases"), when explanations were constructed ("What is my problem? I cannot put in the frequency, etc."), and when the task was structured ("I first want to stabilize the aircraft before finishing the ECAM checklist").

Hypotheses and Results

In the following section we present our hypotheses on predictors of good performance; these hypotheses are based on the considerations outlined above. For each hypothesis, we summarize the results of the field study. Details can be found in Haeusler (2006).

Hypothesis 1.1: adaptive strategies are positively related to performance. Regression analyses summarized in Table 4.2 yield non-significant prediction of subjective overall performance by the degree of adaptation. In contrast, the choice of adaptive task strategies does predict objective technical performance measured by flight parameters. Thus hypothesis 1.1 is confirmed only on the specific level of measurements for adaptation and performance.

Hypothesis 1.2: monitoring/coaching by the pilot non-flying improves performance beyond adaptive task strategies. As shown in Table 4.2, monitoring/coaching does not predict performance beyond adaptive task strategies. Hypothesis 1.2 cannot be confirmed.

Hypothesis 1.3: the universal strategy to augment effort (try harder) does not improve performance. Regression analyses yield only one significant prediction of performance (variability of speed in scenario 2) by the universal strategy to augment effort: Variability of speed is diminished with increased effort. The majority of the analyses confirms hypothesis 1.3 (indicated by framed boxes in Table 4.2).

Hypothesis 1.4: adaptive task strategies moderate the relationship between effort and performance. Higher effort will lead to better performance only in combination with an adaptive task strategy. Three out of nine analyses yielded significant moderator effects that confirm hypothesis 1.4: more effort only goes along with better performance when crews use adaptive task strategies. For instance, there were fewer deviations from the localizer. These deviations were also less prominent. This was achieved through more effort by the pilot flying, but only when he/she shared controls with the pilot non-flying (see Figure 4.2).

Hypothesis 2: adaptive task strategies correspond with less subjective workload. Regression results do not confirm this hypothesis: no relationship between the choice of adaptive performance strategies and subjective workload

Table 4.2   Overview of adaptation and performance (summary of results from regressions)

	Scenario 1: Performance during complex approach			Scenario 2: Performance during loss of control					
	overall subj. per-formance[b]	success of landing	descending below minimum	overall subj. per-formance[b]	deviation from glidepath	deviation from localizer	variability of pitch	variability of speed	variability of sinkrate
adaptive behavior	0			0					
adaptive task performance strategies									
1.1 "lefthand" Circling		+							
1.2 earlier level-off			—						
2.1 delegating some controls					—	–			
2.2 earlier configuration of AC					—	—		—	
2.3 fixed power Setting[a]				+			**+**	+	**+**
Monitoring/Coaching	0	0	0	0	+	+	+	+	+
Effort	+	0	0	0	0	0	0	—	0
Adaptation x Effort	x	x	**X**	**X**	x	**X**	x	**X**	x

Legend:

Summary of the hierarchical regressions for scenario 1 and 2

+ significant positive relation; + non-significant positive relation

— significant negative relation; – non-significant negative relation

X significant interaction; X non-significant interaction

☐ framed results confirm hypothesis; grey shaded fields: not relevant for analysis

[a] low variability in power (SD); [b] self-rated global performance of "pilot flying"

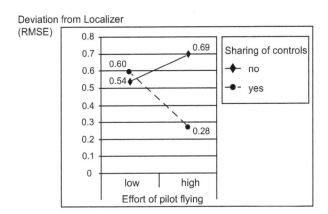

Figure 4.2   Example of moderator effect of adaptation on the relationship between effort and performance

can be found in scenario 1 – the complex approach – whereas a slightly positive relationship was found in scenario 2 – the loss of flight control computers (cf. Haeusler, 2006). The higher the degree of adaptation, the higher the workload participants reported. Adaptation does not necessarily reduce subjective workload, but it can even increase it.

Figure 4.3 depicts the relationship between adaptation and situation awareness, planning, meta-cognitive activity and monitoring/coaching.

Hypothesis 3: cockpit crews demonstrating greater situation awareness (recognizing many demands of the task and the situation) display more planning and strategy development (i.e., they communicate more often about how to cope with the particular task situation in their planning). Bivariate correlations are significant and positive for both scenarios (r = .32; .53) confirming hypothesis 3.

Hypothesis 4: cockpit crews with more extensive planning demonstrate a higher degree of adaptation to situational demands. As expected, more planning goes along with significantly more complete adaptation to situational demands in both scenarios (r = .36; .25).

Hypothesis 5: meta-cognitive activity correlates positively with situation awareness (5.1) and planning (5.2). Figure 4.3 depicts positive correlations between meta-cognition and situation awareness (r = .71; .24) and planning (r = .43; .35). However, hypothesis 5.1 (meta-cognition and situation awareness) is confirmed only in scenario 1 with moderate task load. Hypothesis 5.2 (meta-cognition and planning) is confirmed in both scenarios.

Hypothesis 6: meta-cognitive activity enhances adaptation beyond the influence of situation awareness, planning, and monitoring/coaching. In multiple regressions (see Figure 4.4) meta-cognitive activity predicts adaptation only in scenario 1 (moderate task load). In the high task load scenario with loss of flight control computers, adaptation is best predicted by monitoring/coaching.

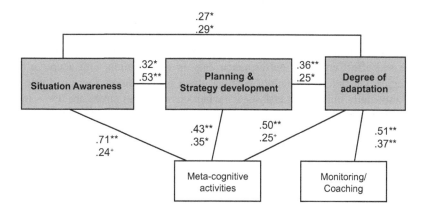

Figure 4.3   **Bivariate correlations (Pearson) of processes influencing adaptation in scenario 1 (bold) and scenario 2 (normal font)**

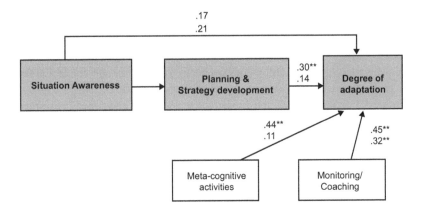

Figure 4.4   Multiple regression predicting adaptation in scenario 1 (bold) and scenario 2 (normal font)

Discussion

Adaptation pays off: cockpit crews who chose adaptive task strategies showed higher technical performance. However, this hypothesis (1.1) was confirmed only with specific measurements for adaptation (for specific adaptive task strategies, but not for the overall degree of adaptation) and for performance (for objective technical performance from flight parameters, but not for subjective overall performance). This might reflect the fact that not all potentially adaptive behavior pays off in a specific scenario. For instance, contingency planning referring to a modified missed approach procedure will not affect performance unless the crew has to fly a go around. Monitoring/coaching was not effective in improving performance beyond adaptation (hypothesis 1.2 not confirmed). The majority of the results confirm that performance does not profit from the universal strategy of augmenting effort in the light of challenges (hypothesis 1.3). Some moderator effects were found, partially confirming hypothesis 1.4, which predicted that increased effort enhances performance only in combination with adaptive task strategies. Despite the generally positive effect of adaptation, there is a price to pay: especially under high task load – when cognitive resources are few – more adaptation goes along with higher subjective workload. This is in line with results from scheduling tasks, where knowledge about the task improved performance at the cost of higher workload (Moray et al., 1991; Segal and Wickens, 1991). This might indicate also that adaptation is best achieved through planning and strategy development during periods of low task load (Orasanu and Salas, 1993). These strategies can then be implemented in high task load situations, during which possibilities for strategy development are likely to be limited. This might also be one reason why adaptation was best predicted by monitoring/coaching in the high task load scenario (see Figure 4.4). This pattern corresponds to a more reactive kind of adaptation that appears when tasks get too complex. For instance, Tulga and Sheridan (1980) found that test subjects were able to develop adequate strategies for moderately complex tasks but tended to use reactive strategies with increasing complexity. To get deeper insights into these questions it would be necessary to use more refined methods to assess a subjective workload than a one-time subjective rating after the event. For a detailed methodological discussion, see Haeusler (2006).

Pilots with more extensive preparation of actions (situation awareness and planning) do adapt better to situational demands (confirming hypotheses 3 and 4). Meta-cognitive activities such as generating hypotheses, observing changes, deducing rules, constructing explanations, and task structuring go together with more elaborate explorations of situational demands (confirmed in scenario 1) and enhance planning (confirmed in both scenarios). The correlations are somewhat stronger in the scenario with moderate task load. Only in the scenario with moderate task load were meta-cognitive activities a significant predictor of adaptation beyond situation awareness, planning and monitoring/coaching. This may again symbolize the trade-off under high task load: cognitive resources are exploited quickly; therefore, exploration, situation awareness and planning are limited. Adaptation under high task load is best predicted by monitoring/coaching (Figure 4.4). Effects of monitoring presuppose the existence of strategies that are already developed and can be retrieved from memory, reducing the task from one of planning to one of comparing.

The difference in the effectiveness of mental preparation (situation awareness and planning) and reflection may also illustrate differences in situational dynamics: with moderate task load, this cognitive effort can be accomplished in the preparatory phase with low task load, since the situational demands can be recognized in advance. In contrast, the more dynamic changes in the high task load scenario cannot be anticipated. There is evidence that under high task load, effortful mental processes such as preparation and planning are abandoned in favor of a more reactive adaptation pattern (e.g., Sauer, Hockey and Wastell, 1999). Another reason for the weaker effects of preparatory and self-reflecting processes under high task load might be due to our assessment of these cognitive processes in terms of communication. High task load may have caused crew members to communicate fewer thoughts.

Practical Implications

Adaptation to situational demands is a mechanism to cope with high task load. To allow flexibility in the choice of task performance strategies, cockpit crews need to be trained in handling a diversity of scenarios that demonstrate operational relevance (e.g., identified by a critical incident reporting system or through analyses based on flight data monitoring). This would lead to a repertoire of rule-

and skill-based options necessary for routine expertise (cf. recognition-primed decision making).

To develop skills that enable pilots to adapt to various situations (adaptive expertise), training needs to go beyond applying learned skills by teaching adaptability (Campbell, 1999). Meta-cognition and self-regulation are important skills: they enhance a person's ability to question situation assessments. In addition, they allow a person to reassess the adequacy of his/her routine behavior and thus facilitate the development of new approaches (Cohen, Freeman and Thompson, 1997). To enhance flexibility and adaptability, trainees should learn to handle classes of activities under varying circumstances (Hacker and Skell, 1993; Kozlowski, 1998). Training should focus mainly on analyzing a situation and on initiating actions. Debriefing should elicit information on the mental models of the pilots. (What aspects of the task and situation did they perceive, how did they judge the situational demands, how did they validate their perception and judgment?)

Results indicate there is room for improvement in cockpit crew training. Training meta-cognitive skills is an investment because it enables pilots to develop more suitable strategies to handle specific tasks. This is a prerequisite for the ability to transfer existing knowledge and skills to novel tasks. Integrating such concepts into new CRM training programs could seriously improve crews' perception of the importance of CRM training. Ultimately, every training organization has to answer the following question: is CRM training simply a means of minimum compliance or can it make a significant contribution in terms of effective crew training with a view to achieving better performance and safety in the cockpit?

Acknowledgements

We would like to thank Captain Werner Naef, former head of CRM training at Swissair, for all his support in realizing the GIHRE project (Group Interaction in High Risk Environments). We also thank Dr Barbara Klampfer and Andrea Welten for the fruitful teamwork in the quest for excellence in the cockpit. We are grateful to Captain Matthias Bosshard and Captain Felix Senn for their patience and skills in explaining technical matters to our team. We thank William Agius for his quest for excellence in correcting the linguistic side of our work.

References

Bargh, J.A. and Gollwitzer, P.M. (1994). Environmental control of goal-directed action: automatic and strategic contingencies between situations and behavior. In: W.D. Spaulding (ed.), *Integrative Views of Motivation, Cognition, and Emotion*. Nebraska symposium on motivation, 41, (pp. 71–124). Lincoln, NE: University of Nebraska Press.

Bateman, T.S. and Crant, J.M. (1993). The proactive component of organizational behavior: a measure and correlates. *Journal of Organizational Behavior*, 14, 103–18.

Berardi-Coletta, B., Buyer, L.S., Dominowski, R.L. and Rellinger, E.R. (1995). Metacognition and problem solving: a process-oriented approach. *Journal of Experimental Psychology: Learning, Memory, and Cognition*, 21, 205–23.

Brannick, M.T., Prince, A., Prince, C. and Salas, E. (1995). The measurement of team process. *Human Factors*, 37(3), 641–51.

Campbell, D.J. (1988). Task complexity: a review and analysis. *Academy of Management Review*, 13, 40–52.

Campbell, D.J. (1991). Goal level, task complexity, and strategy development: a review and analysis. *Human Performance*, 4, 1–31.

Campbell, J.P. (1999). The definition and measurement of performance in the new age. In: D.R. Ilgen and E.D. Pulakos (eds), *The Changing Nature of Performance. Implications for Staffing, Motivation and Development*, (pp. 399–429). San Francisco, CA: Jossey-Bass Publishers.

Cañas, J.J., Quesada, J.F., Adoracion, A. and Inmaculada, F. (2003). Cognitive flexibility and adaptability to environmental changes in dynamic complex problem-solving tasks. *Ergonomics*, 46(5), 482–501.

Cohen, M.S., Freeman, J.T. and Thompson, B.B. (1997). Training the naturalistic decision maker. In: C.E. Zsambok and G. Klein (eds), *Naturalistic Decision Making*, (pp. 257–68). Mahwah, NJ: Erlbaum.

Cohen, M.S., Freeman, J.T. and Wolf, S. (1996). Metarecognition in time-stressed decision making: recognizing, critiquing, and correcting. *Human Factors*, 38(2), 206–19.

Driskell, J.E. and Johnston, J.H. (1998). Stress exposure training. In: J.A. Cannon-Bowers and E. Salas (eds), *Making Decisions Under Stress. Implications for Individual and Team Training*, (pp. 191–218). Washington, DC: American Psychological Association.

Fay, D. and Frese, M. (2001). The concepts of personal initiative (PI): an overview of validity studies. *Human Performance*, 14, 97–124.

Flavell, J. (1979). Metacognition and cognitive monitoring: a new area of psychological inquiry. *American Psychologist*, 34, 906–11.

Ford, J.K., Smith, E.M., Weissbein, D.A., Gully, S.M. and Salas, E. (1998). Relationships of goal orientation, metacognitive activity, and practice: strategies with learning outcomes and transfer. *Journal of Applied Psychology*, 83(2), 218–33.

Frese, M. and Zapf, D. (1994). Action as the core of work psychology: a German approach. In: H.C. Triandis, M.D. Dunette and L.M. Hough (eds), *Handbook of Industrial and Organizational Psychology*, 4, 2nd edn, (pp. 271–340). Palo Alto, CA: Consulting Psychologists Press.

Funk, K. (1991). Cockpit task management: preliminary definitions, normative theory, error taxonomy, and design recommendations. *The International Journal of Aviation Psychology*, 1, 271–85.

Grote, G., Helmreich, R.L., Straeter, O., Haeusler, R., Zala-Mezoe, E. and Sexton, J.B. (2004). Setting the stage: characteristics of organizations, teams and tasks influencing team processes. In: R. Dietrich and T.M. Childress (eds), *Group Interaction in High Risk Environments*, (pp. 111–39). Aldershot: Ashgate.

Hacker, W. and Skell, W. (1993). *Lernen in der Arbeit*. Berlin: Bundesinstitut für Berufsbildung.

Haeusler, R. (2006). Cockpit Crews auf Erfolgskurs – Eine videobasierte Analyse adaptiver Aufgabenstrategien in beanspruchenden Situationen (Unpublished doctoral dissertation). University of Bern, Switzerland.

Haeusler, R., Klampfer, B., Amacher, A. and Naef, W. (2004). Behavioral markers in analyzing team performance of cockpit crews. In: R. Dietrich and T.M. Childress (eds), *Group Interaction in High Risk Environments*, (pp. 25–37). Aldershot: Ashgate.

Hart, S.G. (1989). Crew workload-management strategies: a critical factor in system performance. In: *Proceedings of the Fifth International Symposium on Aviation Psychology*, (pp. 22–27). Columbus, Ohio State University.

Hart, S.G. and Staveland, L.E. (1988). Development of NASA-TLX (Task Load indeX): results of empirical and theoretical research. In: P.A. Hancock and N. Meshkati (eds), *Human Mental Workload*, (pp. 139–83). Amsterdam: Elsevier.

Hart, S.G. and Wickens, C.D. (1990). Workload assessment and prediction. In: H.R. Booher (ed.), *MANPRINT: An Emerging Technology. Advanced Concepts for Integrating People, Machines, and Organizations*, (pp. 257–300). New York: Van Nostrand Reinhold.

Helmreich, R.L., Klinect, J.R. and Wilhelm, J.A. (1999). Models of threat, error and CRM in flight operations. *Proceedings of the Tenth International Symposium on Aviation Psychology*, May 3–6, 1999. Columbus, OH.

Hockey, G.R.J. (1997). Compensatory control in the regulation of human performance under stress and high workload: a cognitive-energetical framework. *Biological Psychology*, 45, 73–93.

Hockey, G.R.J. (2002). Human performance in the working environment. In: P. Warr (ed.), *Psychology at Work*, 5th edn, (pp. 26–50). London: Penguin.

Holyoak, K.J. (1991). Symbolic connectionism: toward third-generation theories of expertise. In: K.A. Ericsson and J. Smith (eds), *Toward a General Theory of Expertise*, (pp. 301–35). Cambridge: Cambridge University Press.

Kanfer, R. and Ackerman, P.L. (1989). Motivation and cognitive abilities: an integrative/aptitude-treatment-interaction approach to skill acquisition. *Journal of Applied Psychology*, 74(4), 657–90.

Klein, G. (1997). The recognition-primed decision (RPD) model: looking back, looking forward. In: C.E. Zsambok and G. Klein (eds), *Naturalistic Decision Making*, (pp. 285–92). Mahwah, NJ: Lawrence Erlbaum Associates.

Kozlowski, S.W.J. (1998). Training and developing adaptive teams: theory, principles, and research. In: J.A. Cannon-Bowers and E. Salas (eds), *Making Decisions Under Stress*, (pp. 115–53). Washington, DC: American Psychological Association.

Kozlowski, S.W.J., Gully, S.M., Nason, E.R. and Smith, E.M. (1999). Developing adaptive teams: a theory of compilation of performance across levels and time. In: D.R. Ilgen and E.D. Pulakos (eds), *The Changing Nature of Performance. Implications for Staffing, Motivation and Development*, (pp. 240–92). San Francisco, CA: Josse-Bass.

LePine, J.A., Colquitt, J.A. and Erez, A. (2000). Adaptability to changing task contexts: effects of general cognitive ability, conscientiousness, and openness to experience. *Personnel Psychology*, 53, 563–93.

Locke, E.A., Shaw, K.N., Saari, L.M. and Latham, G.P. (1981). Goal setting and task performance: 1969–1980. *Psychological Bulletin*, 90, 125–52.

Marks, M.A. and Panzer, F.J. (2004). The influence of team monitoring on team processes and performance. *Human Performance*, 17(1), 25–41.

Maule, A.J. and Hockey, G.R.J. (1993). State, stress, and time pressure. In: O. Svenson and A.J. Maule (eds), *Time Pressure and Stress in Human Judgment and Decision Making*, (pp. 83–101). New York: Plenum Press.

Maynard, D.C. and Hakel, M.D. (1997). Effects of objective and subjective task complexity on performance. *Human Performance*, 10(4), 303–30.

Meijman, T., Mulder, G., Dormolen van, M. and Cremer, R. (1992). Workload of driving examiners: a psychophysiological field study. In: H. Kragt (ed.), *Enhancing Industrial Performance: Experiences of Integrating the Human Factor*, (pp. 245–59). London: Taylor and Francis.

Moray, N., Dessouky, M.I., Kijowski, B.A. and Adapathya, R. (1991). Strategic behavior, workload and performance in task scheduling. *Human Factors*, 33, 607–29.

Orasanu, J. and Salas, E. (1993). Team decision making in complex environments. In: G.A. Klein, J. Orasanu, R. Calderwood and C.E. Zambok (eds), *Decision Making in Action: Models and Methods*, (pp. 327–45). Norwood, NJ: Ablex Publishing Corporation.

Pepitone, D.D., King, T. and Murphy, M. (1988). The role of flight planning in aircrew decision performance. In: *Proceedings of the SAE Aerospace Technology Conference and Exposition*, SAE Technical Paper Series No 881517. Warrendaled: The Society of Automotive Engineers.

Pulakos, E.D., Arad, S., Donovan, M.A. and Plamondon, K.E. (2000). Adaptability in the workplace: development of a taxonomy of adaptive performance. *Journal of Applied Psychology*, 85, 612–24.

Rasmussen, J. (1986). *Information Processing and Human-machine Interaction: An Approach to Cognitive Engineering*. New York: Elsevier Science Publishing Co.

Reason, J.T. (1990). *Human Error*. Cambridge: Cambridge University Press.

Sauer, J., Hockey, G.R.J. and Wastell, D.G. (1999). Maintenance of complex performance during a 135-day spaceflight simulation. *Aviation, Space and Environmental Medicine*, 70(3), 236–44.

Segal, L.D. and Wickens, C.D. (1991). TASKILLAN II: Pilot strategies for workload management. In: *Proceedings of the 34th meeting of the Human Factors Society*, (pp. 66–70). Santa Monica, CA: Human Factors Society.

Shepherd, A. (2001). *Hierarchical Task Analysis*. London: Taylor and Francis.

Smith, E.M., Ford, J.K. and Kozlowski, W.J. (1997). Building adaptive expertise: implications for training design strategies. In: M.A. Quiñones and A. Ehrenstein (eds), *Training for a Rapidly Changing Workplace*, (pp. 89–118). Washington, DC: American Psychological Association.

Sonnentag, S. (2000). Expertise at work: experience and excellent performance. In: C.L. Cooper and I.T. Robertson (eds), *International Review of Industrial and Organizational Psychology*, (pp. 223–64). Chichester: Wiley.

Sperandio, J.-C. (1978). The regulation of working methods as a function of workload among air traffic controllers. *Ergonomics*, 21(3), 195–202.

Sujan, H. (1986). Smarter versus harder: an exploratory attributional analysis of salespeople's motivation. *Journal of Marketing Research*, 23(1), 41–9.

Tripoli, A.M. (1998). Planning and allocating: strategies for managing priorities in complex jobs. *Journal of Work and Organizational Psychology*, 7(4), 455–76.

Tschan, F., Semmer, N.K., Gurtner, A., Bizzari, L., Spychiger, M., Breuer, M. and Marsch, S.U. (2009). Explicit reasoning, confirmation bias, and illusory transactive memory: predicting diagnostic accuracy in medical emergency driven teams in a simulator setting. *Small Group Research*, 40, 271–300.

Tschan, F., Semmer, N.K., Naegele, C. and Gurtner, A. (2000). Task adaptive behavior and performance in groups. *Group Processes and Intergroup Relations*, 3(4), 367–86.

Tulga, M.K. and Sheridan, T.B. (1980). Dynamic decision and workload in multitask supervisory control. *IDDD Transactions on Systems, Man, and Cybernetics*, SMC–10, 217–32.

Waller, M.J. (1999). The timing of adaptive group responses to non-routine events. *Academy of Management Journal*, 42(2), 127–37.

Wood, R.E. (1986). Task complexity: definition of a construct. *Organizational Behavior and Human Decision Processes*, 37, 60–82.

Chapter 5

Manual Flying Skill Decay: Evaluating Objective Performance Measures

Matt Ebbatson, Don Harris, John Huddlestone and Rodney Sears
Cranfield University, United Kingdom

Introduction: The Practical Problem

Despite the capability of modern jet transport aircraft to automatically manage their own flight path and energy for most phases of flight, there are occasions when reversion to basic manual control is essential or even preferable. Pilots may be forced to control the aircraft manually during abnormal situations, such as when recovering from unusual attitudes outside the automations limits of authority or during automation failures. Alternatively, during normal operations pilots may choose to adopt manual control where reconfiguring the automation may not be the most efficient option, for instance when performing a "sidestep" (US) manoeuvre to accommodate a late change in the assigned arrival runway. However, during the vast majority of airline operations the requirement for manual flight is limited. For a typical sector it may extend to just a few minutes following takeoff and a short period immediately before landing.

Every licensed air transport pilot must demonstrate their ability to operate the aircraft manually during recurrent training and proficiency checks in full flight simulators. However, simulator time is valuable to an airline. There are numerous items in addition to manual flying ability which must be assessed during these brief sessions. For example, the requirement for pilots to validate their low visibility operations proficiency dictates that much of the session may be flown using the automatics. Consequently the amount of time dedicated to manual handling may be minimal.

In March 2002 the crew of an Airbus A321 returning to East Midlands Airport in the UK elected to perform a manual approach, disengaging both the autopilot and the auto thrust systems. Despite good visibility the handling pilot selected raw ILS data to provide him with additional guidance. The crew failed to arm the flight director approach mode and consequently the aircraft became slightly displaced above the glideslope. The flight director was immediately disengaged but the handling pilot failed to properly manually recover the glideslope until late in the approach, deviating substantially both above and below it. In addition, the aircraft also deviated significantly on the localiser, both to the left and to the right. During the final stages of the approach, the profile was recovered but the thrust was improperly managed and the airspeed decayed well below the target speed. While flaring the aircraft, an abnormally large angle of attack was required to reduce the descent rate due to the low airspeed and, consequently, the tail struck the runway causing damage to the fuselage skin and supporting structure. The handling pilot had been operating the aircraft type for just over a year but previously only had extensive experience operating single seat military fast jets. He later commented that he could not recall having ever operated his current aircraft without the use of auto thrust before the incident date (AAIB, 2002).

In the bulletins issued by the UK Air Accident Investigation Branch (AAIB) there are accounts of at least two very similar incidents occurring relatively recently. On both occasions, the autopilot, auto thrust and flight director systems were disengaged at an early stage of the approach, as in the East Midlands incident. Also on both occasions, there were significant deviations on the localizer and glideslope, followed by poor airspeed management in the latter stages of the approach resulting in high flare angles and tail strike damage. The handling pilot on one occasion was a relatively inexperienced first officer, whilst on the other occasion the handling pilot was a very experienced captain with over 3,000 hours on type. In all the events described, the handling pilot was about to undergo a proficiency assessment and stated that they had deliberately flown the aircraft manually to practice handling the aircraft prior to their evaluation.

These incidents are symptomatic of a wider, systemic issue in airline operations. During their initial training all pilots are taught the complex psychomotor and cognitive skills required to control their aircraft by the physical manipulation of the primary flying controls – that is, basic manual flying skills. Conversely, during routine operation of a modern jet transport aircraft it is more common for the continuous flight path control task to be managed by a combination of automated systems (i.e., the autopilot and auto-throttle). In this mode of operation the psycho-motor aspect of control is minimal and the cognitive aspect is modified (Damos et al., 2005; Latorella et al., 2001) with the emphasis of the pilot's work shifting towards higher order cognition (i.e., decision making and problem solving). The opportunity to practise basic manual flight is usually minimal, although this is also somewhat dependent on the type of carrier, its equipment, operational philosophy and route network. For example, two of the most prominent low cost carriers in the UK have distinct operational philosophies, one encouraging routine manual flight over use of the automatics and the other the exact opposite.

However, as noted earlier, the ability of the pilot to revert to basic manual control is essential. Wood's (2004) study of flight crews' dependency on automation noted that strong anecdotal evidence existed to suggest that pilots of highly automated aircraft may experience manual flying skills decay as a result of a lack of opportunity to practice during line operations. (See also Curry, 1985; Veillette, 1995; Owen and Funk, 1997.) The infrequent opportunity to exercise manual flying skills during modern flight operations may cause crews to experience "out-of-the-loop unfamiliarity" (Wickens, 2000) and their basic flying abilities may diminish over time. The threat of this skill fade is a concern shared by pilots, operators, regulators, manufacturers and researchers alike (Baron, 1988; Childs and Spears, 1986; Parasuraman et al., 1993; Tenney et al., 1998). It was suggested that this may pose a threat to flight safety, predominantly during periods of imposed manual flying, such as following a partial degradation (or even outright failure) of the aircraft's automation.

The Aircraft Control Problem

Manual aircraft control requires the pilot to employ both open-loop and closed-loop control behavior (Baron, 1988). Open-loop behavior is independent of feedback, similar to a golfer driving a ball, and involves the execution of pre-programmed motor schema to effect large changes in the aircraft's orientation, path or location. Closed-loop control is used to track and maintain a target state by monitoring feedback channels. In this control mode the pilot monitors and adjusts performance in order to reduce any discrepancy between the desired aircraft state and the observed aircraft state. Feedback is delivered primarily via the flight instrumentation and the outside field of view, although vestibular, somatic, proprioceptive and auditory cues are also utilised. The continuous closed-loop control requirement of manual flight is therefore highly demanding of the pilot's physical and cognitive capacity.

However, it is somewhat simplistic to describe pilot control as falling into these two categories. McRuer and Jex (1967) in their "successive organisation of perception" model proposed that skilled operators could potentially operate in three modes (the highest of which could only be achieved with extensive practise). They suggested that novices operated at the lowest level, which was a pure compensatory control strategy, simply cancelling any error in their desired path. However, this results in relatively poor control performance, as the response times of the human operator, control lags and display lags will all play a part. With increasing practise it becomes possible to operate in a pursuit mode, responding directly to the control input, thereby improving performance by avoiding many of the lags in the system. At the highest level, the operator is responding in a pre-cognitive mode. This is an entirely open-loop behavior that operates independently of any feedback. A highly skilled pilot can exhibit all of these psychomotor skill modes. A large aircraft's flight path always shows an appreciable lag behind a pilot's control inputs, thus, operating in a pre-cognitive mode allows the pilot to minimize or avoid many of the lags inherent in the aircraft's flight control system. However, a pilot's reactions to the disturbing effects of turbulence must be compensatory as the effects of turbulence on flight path cannot be predicted. The experienced pilot compensating for a side-wind on final approach will exhibit pursuit tracking behavior. Although the accuracy of the rudder application will become apparent as the pilot continues the approach, initially the pilot will put in an amount of rudder which "feels about right". Rotating the aircraft to the appropriate pitch attitude on take-off or in the flare, may be considered to be examples of the pre-cognitive mode of control behavior.

The control of an aircraft is best described as a hierarchical control problem. The basic nature of a hierarchical control problem is that the parameter that needs to be controlled (in the case of an aircraft, its flight path) can only be controlled indirectly via other lower order parameters (e.g., pitch rate or roll rate/heading – see Figure 5.1). There is no method of directly controlling an aircraft's altitude. This is accomplished via a pitch rate control, which changes pitch attitude, causing the aircraft to climb. In terms of changing the aircraft's heading, this is done by rolling the aircraft to the appropriate roll attitude, which changes its rate of change of heading, which subsequently changes the heading of the aircraft.

One of the most common control models is the series pilot model, which describes the nature of the pilot/aircraft system as a whole (McRuer, 1982), see Figure 5.2. In this model the control of an aircraft is characterized as a compensatory tracking task in which the objective is to minimize any errors from the desired three-dimensional flight path. However, as noted earlier, the one thing a pilot cannot directly control is flight path, which is accomplished via a series of lower order surrogates. For example, in the vertical axis the pilot controls pitch rate in the short term or angle of attack in the longer term (see Field, 2004). However, there is also one further problem. The pilot cannot actually observe either flight path or angle of attack directly; these can only be inferred from the aircraft's pitch attitude. In the series pilot model, the flight control problem is decomposed into a short-term and a long-term problem. In the vertical axis, the short-term (or inner loop) control problem is one of pitch attitude control because the pilot cannot directly observe angle of attack, which is the parameter that has an actual effect on flight path. The longer-term (outer loop) control problem is one of flight path or altitude control.

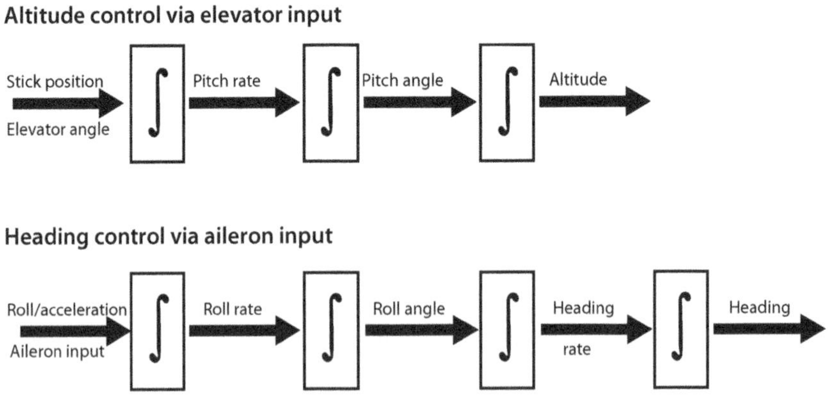

Altitude control via elevator input

Heading control via aileron input

Figure 5.1   The hierarchical nature of the aircraft control problem

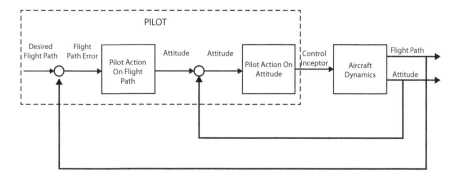

Figure 5.2   **The series model of pilot control (adapted from McRuer, 1982)**

The short term (pitch attitude) control problem can be considered to be "nested" within the longer term (flight path angle/altitude) control problem. What the pilot is doing is attempting to "close" the inner control loop (the attitude control problem) in an attempt to control the "real" problem, which is that of altitude/flight path angle. In the horizontal plane, the pilot is dealing with a similar control problem, only in this case aircraft roll attitude is being used as a first approximation (inner loop parameter) in an attempt to control the aircraft's heading and, subsequently, its track across the face of the earth.

The Measurement Problem

Flight Path Derived Measures

When evaluating performance on any tracking task, such as flying an aircraft, it is common to examine the end product of performance (i.e., measuring errors between the tracked parameter and a target value). Metrics, such as the arithmetic mean error and standard deviation of error, have strong validity when applied to parameters such as flight path or airspeed deviation especially when associated with a well-prescribed flight task that demands a high level of performance, for instance, flying an ILS-approach. The arithmetic mean error gives an indication of the overall flight path error (on a particular axis) and its associated standard deviation gives a measure of the "smoothness" of the pilot's performance. These two parameters are often used in preference to the Root Mean Square Error (RMSE). Taken in combination, the arithmetic mean error and the standard deviation of error completely define the root mean square error. Furthermore RMSE also has the additional disadvantage that it produces identical values for quite disparate performances. For example, being consistently high, consistently low, or at the correct mean height but with great variations in height-keeping, may all result in the same RMSE value (see Hubbard, 1987). Measures such as the arithmetic mean error and the standard deviation of error have been used on many occasions to evaluate pilot performance. All measures that relate to the flight path of the aircraft can be considered to reflect the outer loop performance in McRuer's series pilot model.

Unfortunately, in a large conventional transport aircraft, the relationship between control input, aircraft attitude and flight path variation is mediated by factors such as inertia, control power and the relatively high stability of the machine. Unlike smaller aircraft, there is often a significant delay between control input and the larger aircraft's response. Consequently, further control inputs after the initial input may serve to cancel it out or reinforce the initial input before it has taken effect. As a result, significant control input activity may not be reflected in large changes in the aircraft's attitude, and less so in its flight path. As a result, the pilot's control inputs (inner control loop parameters relating to changes in aircraft attitude) will not necessarily be reflected for some considerable time in the outer control loop, flight path-based parameters. Consequently, two pilots may describe similar flight paths but control the aircraft in a very different manner. Simple measurement of flight path error can determine if a pilot's control actions were ultimately successful, but these may only be a crude measure of the process by which the result was achieved. It is, therefore, unlikely that basic flight path measures alone will have the sensitivity required to investigate fine variations in manual flying skills. However, there is the potential for a direct assessment of the pilot's control strategy. A pilot's control strategy can be assessed directly by studying his inputs to the primary flight controls as an adjunct to the aforementioned flight path metrics. The assessment of pilots' control inputs to the primary flight controls provides a direct measure of their behaviour associated with closing the inner (attitude) control loop.

Control Input Derived Measures

McDowell (1978) developed control movement power spectra-based measures (a description of the distribution and weighting of control input frequencies and amplitudes) to evaluate pilot performance. These measures relate directly to the magnitude and frequency of the pilots' inputs to the flight controls *not* to the flight path of the aircraft.

The power spectrum shows how the power of a control input (energy per unit time) is distributed over a frequency range and, thus, by examination, it becomes possible to determine how much of the control input power falls into a given frequency band. This is essentially the approach used by McDowell. The typical means of computing the power spectrum is by performing a discreet Fourier transform. The Discreet Fourier Transform (DFT) identifies periodicities in a series of measured data and measures the relative strength of that periodicity (Press, Flannery, Teukolsky and Vetterling, 1989). Such algorithms work on the assumption that any complex waveform can be expanded into a superposition of sines and cosines of varying amplitude, frequency and phase (see Figure 5.3). The discreet Fourier transform, Fn, of a series of data fk with N data points is given by:

$$F_n \equiv \sum_{k=0}^{N-1} f_k e^{-2\pi i n k / N}$$

With appropriate scaling, the coefficients of the DFT give the power spectral density (PSD) of the time series data (see Figure 5.4) expressing power per unit frequency. Peaks in the distribution show strong periodicities in the signal and indicate where power is concentrated. Although scaling and interpretation of this process can be complex, many maths processing packages such as Matlab™ incorporate streamlined DFT functions. By integrating the PSD data between two frequency limits, it is possible to determine the amount of control input power within that frequency band.

Using this approach, it was observed that more experienced pilots flying a Cessna T-37 light military training aircraft generally used higher frequency control inputs, particularly in the roll axis. McDowell concluded that there were changes in pilot's control movement power spectra as a function of skill level, and that measures of this property could be used effectively to discriminate pilot skill/experience level. However, the control input data obtained and subsequent analyses are specific to the type of aircraft, the task and the environment in which it is conducted.

Johnson, Rantanen and Talleur (2004) developed several new metrics to quantify differences in PSD distributions which were associated with performance. These are summarized in Table 5.1. These metrics described things such as the average amplitude of significant components and their spread in amplitude (the mean magnitude of spectral components – MSC – and the standard deviation of the magnitude of spectral components – DSC – respectively). It was demonstrated that these metrics were capable of discriminating between pilots who had passed or failed an instrument flying proficiency check. However, in this case the metrics were applied primarily to flight path (outer loop) parameters, such as course deviation and glideslope deviation indications, rather than directly to primary flight control movement data.

Ebbatson and co-workers, in a series of publications, have demonstrated the utility of using frequency-based measures for assessing pilot performance when flying heavy, civil aircraft. Ebbatson, Huddlestone, Harris and Sears (2006) evaluated the performance of 12 cadet pilots while undertaking a 40-hour Jet Orientation course on a Boeing 737NG. The results showed that variation in the flight path was reduced as the cadet pilots progressed through the course. However, at the later stages of the course, the control strategy used changed and was characterized by more frequent but smaller amplitude control inputs. Over the

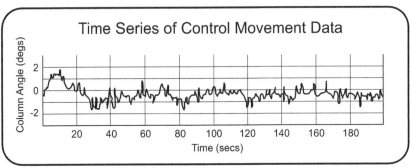

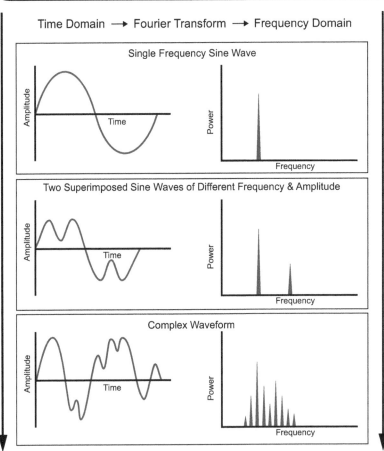

Figure 5.3   The transposition of time series data into the frequency domain using a fast Fourier transform algorithm

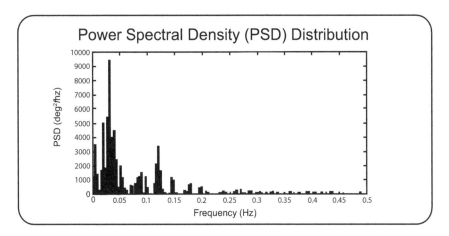

Figure 5.4 **Example plot of power spectral density distribution (periodogram) for a control input**

Table 5.1 **Fourier analysis based performance metrics as developed by Rantanen et al. (2004)**

Metric	Description
Mean Magnitude of Spectral Components(MSC)	Mean of the $\dfrac{\left\|\widetilde{Y}_j\right\|^2}{N}$, relating to the amplitude of deviations in the time series
Number of Spectral Component Greater than Cut-Off (NCGC)	Number of $\dfrac{\left\|\widetilde{Y}_j\right\|^2}{N}$ greater than a minimum cut-off value to remove noise
Mean Frequency of Spectral Components Greater than Cut-Off (FMGC)	Mean frequency of the $\dfrac{\left\|\widetilde{Y}_j\right\|^2}{N}$ greater than a minimum cut-off value
Standard Deviation of the Frequency of Spectral Components Greater than Cut-Off (FDGC)	SD of frequencies of the $\dfrac{\left\|\widetilde{Y}_j\right\|^2}{N}$ greater than a minimum cut-off value
Median Frequency of the Spectral Components (MEDF)	Median frequency of the power spectrum

period of the training course there was an increase in the mean pitch input frequency in the spectral components, as well as an increase in the central frequency of the spectral distribution. In addition, the mean Power Spectral Density decreased over this period, indicating a decrease in the amplitude of pitch inputs.

The utility of frequency-based measures over flight path derived metrics was further demonstrated when the performance of the same pilots was evaluated during a series of symmetric and asymmetric approaches undertaken during their training program (Ebbatson, Harris, Huddlestone and Sears, 2008). The flight path data were analyzed for deviations around the optimum flight path while flying an instrument landing system (ILS) approach. Data from the manipulation of the flight controls were subject to analysis using a series of power spectral density measures. The results demonstrated that when assessing pilot performance flying a large jet transport aircraft, the approach of measuring performance from errors in flight path was insensitive, even in an extreme situation such as when undertaking a single-engined approach. Despite the asymmetric thrust condition being significantly more demanding, the traditional flight path error metrics indicated that there was very little difference between pilots' performance in the two conditions. The mean error metric showed that the pilots tracked equally close to the glideslope and localizer in the asymmetric condition, whilst the standard deviation of error metric suggested that there was also no difference in the smoothness of tracking those targets.

However, when examining the control inputs used to achieve what appeared to be similar performances in ILS tracking, it was revealed that very different control strategies were employed between the symmetric and asymmetric approaches. During asymmetric approaches roll inputs were made with far greater amplitude and over a wider but generally lower band of frequencies. Similarly, yaw control magnitudes were much larger (although at much the same frequencies). It was not very surprising that both control wheel and rudder inputs were of different magnitudes in the two scenarios given the lateral disturbance the asymmetric thrust condition generated. However, the significant differences in the pilots' performance could only be determined by the manner in which they controlled the aircraft and not from simply investigating the relatively coarse, flight path derived measures of outer loop performance.

In response to the concern that pilots of highly automated aircraft experience a decay in their manual flying skills as a result of a lack of opportunity to practise hand-flying during line operations, Ebbatson, Harris, Huddlestone and Sears (2010) examined the relationship between pilots' manual handling performance and their recent flying experience using both traditional flight path tracking measures and frequency-based control strategy measures. Measures were taken during several flight segments from 66 professional airline pilots during the crew's biannual License Proficiency Check. All pilots held an Air Transport Pilot License and a Boeing 737-300/400/500 type endorsement. Prior to undertaking the trials, pilots were asked to provide details about the number of sectors they had flown during the previous month. Participants were each required to perform a standardized terminal manoeuvring exercise in instrument meteorological conditions again with the aircraft in an asymmetric thrust condition. The exercise incorporated a variety of demanding, but operationally relevant manual flight tasks with the autopilot, flight director and auto throttle systems all inoperative. The task began from a straight and level condition at platform altitude with the aircraft at relatively high speed in a clean configuration (flaps and gear not deployed) positioned to intercept the ILS localizer. Whilst performing the single engine ILS component of the task, a significant backing crosswind further added to the demands on the pilots. As weather conditions at decision height prevented visual acquisition of the runway, this required that a single-engine missed approach procedure be performed which terminated the exercise.

Significant relationships were identified between pilots' recent flying experience and their manual control strategy. The results showed that flight path derived measures were again relatively insensitive to the amount of recent flying undertaken by the pilots. In the ILS segment of flight, only the glideslope standard deviation of error showed any significant correlation with recent manual flying experience: in the straight and level segment, only the heading error exhibited a significant relationship. However, the inner-loop parameters proved to be much more sensitive to recent flight experience, with several frequency-based measures showing a significant association. During the straight and level and ILS flight segments, the analyses demonstrated that pilots who had flown more sectors in the previous week tended to use lower frequency control inputs in pitch and exhibited a narrower spread of pitch input frequencies to the control column. Similarly the data indicated that pilots who had flown more sectors in the previous week used lower frequency yaw control inputs and exhibited a narrower spread of input frequencies to the rudder.

Conclusions

In certain conditions traditional performance metrics based on flight path tracking error may be insensitive and fail to identify important differences in the pilot's performance. It is argued that the use of frequency-based measures of control inputs are a necessary adjunct for evaluating pilot performance when flying large jet transport aircraft with high inertia and numerous lags in the control system, which makes the aircraft much less agile and slower to respond. These metrics give an extra dimension to performance measurement, relating directly to the inner loop of the series control model and describing the process by which pilots exercise their control. These measures are far more sensitive than flight path derived measures, as demonstrated across a range of applications, and enable a more complete picture to be developed for the study of manual flying skills.

References

AAIB (2002). Airbus A321-211 G-JSJX, AAIB Bulletin no. 3/2002, Air Accident Investigation Branch, Farnborough, Hants.

Baron, S. (1988). Pilot Control. In: E.L. Wiener and D.C. Nagel (eds). *Human Factors in Aviation*, (pp. 347–385). San Diego, CA: Academic Press.

Childs, J. and Spears, W. (1986). Flight-skill decay and recurrent training. *Perceptual and Motor Skills*, 62, 235–42.

Curry, R.E. (1985). The introduction of new cockpit technology: a human factors study. NASA Technical Memorandum 86659, 1–68. Moffett Field, CA: NASA Ames Research Center.

Damos, D., John, R. and Lyall, E. (2005). Pilot activities and the level of cockpit automation. *International Journal of Aviation Psychology*, 15(3), 251–68.

Ebbatson, M., Harris, D., Huddlestone, J. and Sears, R. (2008). Combining control input with flight path data to evaluate pilot performance in transport aircraft. *Aviation Space and Environmental Medicine*, 79(11), 1061–4.

Ebbatson, M., Harris, D., Huddlestone, J. and Sears, R. (2010). The relationship between manual handling performance and recent flying experience in air transport pilots. *Ergonomics*, 53(2), 268–77.

Ebbatson, M., Huddlestone, J., Harris, D. and Sears, R. (2006). The application of frequency analysis based performance measures as an adjunct to flight path derived measures of pilot performance. *Human Factors and Aerospace Safety*, 6(4), 383–94.

Field, E. (2004). Handling qualities and their implications for flight deck design. In: D. Harris (ed.), *Human Factors for Civil Flight Deck Design*, (pp. 157–81). Aldershot: Ashgate.

Hubbard, D.C. (1987). Inadequacy of root mean square error as a performance measure. In: R.S. Jensen (ed.), *Proceedings of the Fourth International Symposium on Aviation Psychology*, (pp. 698–704). Columbus, OH: Ohio State University.

Johnson, N.R., Rantanen, E.M. and Talleur, D.A. (2004). Criterion setting for objective fourier analysis based pilot performance metrics. *Proceedings of the Human Factors and Ergonomics Society 48th Annual Meeting, Santa Monica*.

Latorella, K., Pliske, R., Hutton, R. and Chrenka, J. (2001). Cognitive task analysis of business jet pilots' weather flying behaviours: preliminary results. NASA TM 211034, NASA Langley.

McDowell, E.D. (1978). *The development and evaluation of objective frequency domain based pilot performance measures in ASUPT*. Air Force Office of Scientific Research, Bollings AFB, DC.

McRuer, D.T. (1982). Pitfalls and progress in advanced flight control systems (AGARD CP-321). Neulliy-sur-Seine: AGARD/NATO.

McRuer, D.T. and Jex, H.R. (1967). A review of quasi-linear pilot models. *IEEE Transactions on Human Factors in Electronics, HFE-8*, 3, 231–49.

Owen, G. and Funk, K. (1997). Flight deck automation issues: incident report analysis. <http://www.flightdeckautomation.com/incidentstudy/incidentanalysis.aspx> (accessed September 1, 2011). Corvallis, OR: Oregon State University, Department of Industrial and Manufacturing Engineering.

Parasuraman, R., Molloy., R. and Singh, I. (1993). Performance consequences of automation-induced "complacency". *International Journal of Aviation Psychology*, 3(1), 1–23.

Press, W., Flannery, B., Teukolsky, S. and Vetterling, W. (1989). Fourier transform of discretely sampled data. *Numerical Recipes in FORTRAN: The Art of Scientific Computing*, 2nd ed. Cambridge: Cambridge University Press.

Tenney, Y., Rogers, W. and Pew, R. (1998). Pilot opinion on cockpit automation issues. *International Journal of Aviation Psychology*, 8(2), 103–20.

Veillette, P. (1995). Differences in aircrew manual skills in automated and conventional flight decks. *Transport Research Record 1480*, 43–50.

Wickens, C.D. (2000). *Aviation Psychology*, Oxford: Oxford University Press.

Wood, S. (2004). Flight crew reliance on automation. CAA Report 2004/10. Civil Aviation Authority, Gatwick.

Chapter 6

Civil Pilots' Stress and Coping Behaviors: A Comparison Between Taiwanese and Non-Taiwanese Aviators

Chian-Fang G. Cherng
Chang Jung Christian University, Taiwan, ROC

Jian Shiu
Civil Aeronautics Administration, Taiwan, ROC

Te-Sheng Wen
Mingdao University, Taiwan, ROC

Introduction

On evaluating the occupational stress for 104 jobs, Cooper et al. (1988) demonstrated that civil aviation (pilot) was rated as one of the extremely stressful jobs. A previous study has documented that job stress, mainly related to work accidents or reduced job performance (DuBrin, 2004) is recognized as a major risk to workers' well-being (Kushnir, 1995). Since pilots' work is categorized as high stress, it is of importance to understand civil pilots' baseline stress level, sources of stress, their stress reactions and habitual coping behaviors. Flight safety can be enhanced by establishing pilot selection criteria and reinforcing educational and training programs accordingly.

Stress is a Subjective Issue

Stress refers to a set of changes that people undergo in situations that they appraise as threatening to their well-being (Auerbach and Gramling, 1998). These changes involve physiological arousal, subjective feelings of discomfort, and overt behavior. In fact, Zuckerman (1991) claimed that one person's stress can be another person's pleasure. That is, stress is partly a matter of individual judgment or appraisal and is individual-dependent in nature (Thom, 1997).

Even more so, Green et al. (1996) further referred to stress as a person's evaluation downgrading his/her own ability to meet the perceived demands. All evidence considered, stress is subjective rather than objective. Li et al. (2001) analyzed a set of 29891 data related to aviation crashes for the years of 1983–1996 and found that pilot error was a probable the cause in 85 percent of aviation crashes. They therefore suggested that flight safety was the balance between performance ability and performance demand. Interestingly, they explained the weather-related crash and crash location by using Selye's stress model (1974): reaction to stressors was processed by three stages, the alarm stage, the resistance stage, and the exhaustion stage, respectively. Li et al. (2001) posited that a stressor could damage flight safety by either increasing performance demand or reducing pilots' performance ability. A crash, then, occurred at the exhaustion stage when performance demand exceeded performance ability. That is, flight accidents may be highly associated with aviators' subjective stress.

Stressors

The circumstances or situations that induce stress responses are defined as stressors (Auerbach and Gramling, 1998). Green et al. (1996) outlined four types of pilots' stressors, these are environmental stressors, domestic stress, work stress, and life stress. On the flight deck, environmental stressors specifically consist of excessive heat, noise, vibration, and low humidity.

Even domestic stress can affect pilots' concentration and performance while flying an aircraft. For example, numerous accident and incident studies showed that an error of skill could be made by a pilot due to his/her pre-occupation with his/her domestic problems (Alkov et al., 1985). Any change in a person's domestic situation or relationship, such as divorce, marital separation, new baby, poor relationship with partner or other family members, can be a stressor (Thom, 1997). The loss of a spouse has been found to lead to the highest level of stress in all kinds of life stress experience (Holmes and Rahe, 1967). A pilot, separated from his family by a long distance and concerned about urgent matters or financial problems, can find himself or herself under devastating stress.

Furthermore, work stressors for pilots can also arise from the job demands or workloads, emergency reactions, operating environment, company policies, and relationships with colleagues. Pressure and conflict in work operations could be potential stressors. For example flight companies' policies, based on financial considerations that require pilots to operate tight flying schedules with insufficient rest, can inevitably lead to seriously jeopardized flight safety. It is of importance to note that a company's policies imposed by management are not only restricted to the flying roster but may also affect promotion, career development, etc. Finally, poor communication and conflict in relationships with colleagues represent yet another source of stressors. These events and conditions are all potential sources of stress at work.

In fact, all sorts of life changes can be considered to be stressful. It has been found that the effects of an individual's life changes tend to be cumulative and are very likely linked to illness (Holmes and Rahe, 1967; Selye, 1974). For example, pilots with insufficient flying duties may worry about their job security, whereas pilots with overloaded flying missions may worry about their professional efficiency and domestic problems (Green et al., 1996).

Stress Reactions

The possible reactions in response to a subjective stress can be classified into three main domains, psychological, physiological, and behavioral consequences. The *psychological reactions* can be subdivided into two aspects, one is cognition-related and the other is emotion-related. *Cognitive reactions* include lack of concentration, forgetfulness, inability to make good decisions, etc. *Emotional reactions* are those related to feelings of anxiety, depression, tension, irritability, etc.

Physiological reactions refer to the symptoms associated with stress, e.g., headaches, loss of energy, etc. Physiological symptoms are associated with high risk of health problems, such as coronary heart diseases. *Behavioral reactions* include performance impairment, sluggish reaction time to make responses, etc.

People suffering from stress are prone to accidents (Green et al., 1996). In consideration for flight safety pilots suffering stress should be carefully supervised or assisted.

Stress Coping Types

In Aldwin and Revenson's definition (1987), coping encompasses cognitive and behavioral strategies used to manage a stressful situation (problem-focused coping) and the attendant negative emotions (emotion-focused coping). Coping is the process whereby the individual either adjusts to the perceived demands of the situation or changes the situation itself. Green et al. (1996) categorized coping strategies into three domains: action coping, cognitive coping, and symptom-directed coping. Using action coping, the individual attempts to reduce the stress by taking action, either to remove the problem or alter the situation. By taking cognitive coping, the individual reduces the emotional and physiological impact of stress on himself or herself. Symptom-directed coping involves the use of drugs, tobacco, alcohol, tea or coffee. Physical exercise, meditation, and other stress management techniques can be effective, to some extent, for this type of coping.

Havlovic and Keenan (1995) divided coping behaviors into five types: *Positive thinking*. For example, "think of ways to use this situation to show what one can do or what can be done?" or "try to see this situation as an opportunity to learn and develop new skills." *Direct action*. For example, "devote more time and energy to doing jobs" or "try to work harder and more efficiently". *Help-seeking*. For instance, "request help from other people who have the power to do something" or "talk with people who are involved". *Avoidance/Resignation*. For instance, "try to keep away from this type of situation" or "avoid being in this situation as much as possible". *Use of substances.* For example, "drink beer or wine" or "smoke a cigarette or cigar."

The first three types were the control orientated adaptive coping activities and the remaining two types were maladaptive coping methods. A number of variables, including personality, job tenure, and managerial experience, have been found to be associated with specific coping types. As shown in Havlovic and Keenan's results (1995), Type A personalities, those who with greater job tenure and managerial experience, were more likely to use positive thinking coping activities.

Culture and Flight Safety

People in different countries may hold different values within their lifestyles for behavior or work. Cumulative evidence reveals that there are substantial differences in the way pilots of different nationalities conduct their tasks in the cockpit. These differences are mainly derived from their cultures and the cultural differences certainly have clear implications for safety (Helmreich and Merritt, 1998). As shown in their study, strong cross-cultural differences in the areas of communications and tolerance for rules, routines and set procedures were observed. These differences of pilots' conduct in the cockpit can, in turn, contribute to an increase of uncertainty and hesitation and therefore become a serious threat to safety (Helmreich and Merritt, 1998).

In an attempt to emphasize national value differences, Hofstede (1980) employs a four-dimension category to discriminate among various national cultures. They are power distance (PD), uncertainty avoidance (UA), individualism/collectivism, and masculinity/femininity, respectively. PD is the extent to which the less powerful members of organizations accept and expect that power is distributed unequally. In a low PD culture, subordinates feel more comfortable approaching or contradicting their superiors. In contrast, there is considerable dependence of subordinates on superiors, and subordinates are unlikely to question their superiors directly in a high PD culture (Helmreich and Merritt, 1998).

UA deals with a society's tolerance for uncertainty and ambiguity. It indicates to what extent a culture programs its members to feel either uncomfortable or comfortable in unstructured situations. Hofstede (1991) argues that high UA countries could have more people who feel under stress at work, want rules to be respected, and want to have a long-term stable career. People in high UA cultures are more driven to keep busy and are often more precise, whereas people in low UA countries are more relaxed in their work.

Individualism/collectivism is the degree to which individuals are integrated into groups. In the individualist societies, the ties between individuals are loose. In contrast, people are integrated into strong, cohesive in-groups in the collectivist societies. Individualists consider the implications of their behavior within a narrowly defined area of personal costs and benefits. Independence and self-sufficiency are valued, with individual achievement and recognition being preferred to group-based rewards (Spence, 1985). Self-reliance is a strength, while seeking for help implies weakness and mistakes (Hui and Triandis, 1986). In contrast, people in collectivist cultures consider the implications of their behavior in linking to their in-groups, such as family or organization. They value loyalty and harmony within the group.

Masculinity/femininity refers to the distribution of emotional roles between genders. Men are supposed to be assertive, tough and focused on extrinsic achievement, e.g., high earnings, promotions, material success. In contrast, women are supposed to be modest, tender and concerned with quality of life. Femininity pertains to societies in which social gender roles overlap, and there is greater concern for quality of life (Helmreich and Merritt, 1998). The men in feminine countries, i.e., countries where women are more or less accepted as equals to men, have the same modest, caring values as the women; in the masculine countries women are more assertive and more competitive, but not as much as the men, so that these countries show a gap between men's and women's values.

The Main Goals of the Present Study

Prior to 1987, most of the local civil pilots in Taiwan were those who retired from their full-time military service. Only one semi-official company hired a few foreign pilots for its international operations. The Taiwanese government claimed the Open Sky policy in 1987. According to the policy, new airline companies were allowed to take part in both domestic and international operations. Under these favorable conditions, the number of local airline companies grew rapidly, and this raised the demand for recruiting more civil pilots. China Airlines started to recruit its own pilot trainees. At the same time, the Civil Aeronautics Administration (CAA) in Taiwan revised the regulation and allowed airline companies to hire more foreign pilots. A new airline company, EVA Airways, setting up business in 1991, hired hundreds of foreign pilots to fly its Boeing jets and started to recruit its own pilot trainees.

Since there is no civil flight training school in Taiwan, local airline companies have to send their pilot trainees overseas for flight training. In US flight training schools, most of the flight cadets attend their *ab-initio* and civil pilot training courses for approximately 15 to 20 months in total. As soon as the training program in flight training school has been completed, these trainees have to go through a six-month evaluation in Taiwan. After successfully passing this evaluation, these trainees can finally start their flight careers and serve as first officer *ab initio*. Based on the statistics published by the CAA (2010), there were 2037 active civil pilots in Taiwan by the year of 2000. Among them, 526 were foreign pilots, and 1511 were Taiwanese pilots (40 percent were civilian-trained pilots and 60 percent were former military pilots). Although the total number of foreign pilots has been progressively declining since 2007 (382 in 2007; 276 in 2008; 202 in 2009), the present airline companies in Taiwan are undoubtedly highly multi-cultured organizations.

As shown by Helmreich and Merritt (1998), cross-cultural differences and the difference-induced problems were expected to occur in the areas of communication and interaction, tolerance for rules, routines and set procedures, especially since local and foreign pilots worked together in the cockpit. Pilots raised and educated in different national cultures, Taiwanese and foreign pilots in this case, may perceive different stress profiles and/or exhibit different stress reactions and stress coping behaviors in response to similar stress types. On the other hand, it was also likely that Taiwanese and foreign pilots did not experience such cultural differences or related problems at all simply because both local and foreign pilots performed their cockpit jobs with a much more salient emphasis on professional culture than on his/her personal culture. Given local and foreign pilots do display differences in primary stressors, stress reaction, and coping profiles, it was reasonable to suspect the occurrence of conflicts and incongruities in the cockpit. In addition, to reduce intra-pilot and pilot-machine conflicts and incongruities in the cockpit, was of similar importance to reduce inter-pilot conflicts and incongruities in order to maintain flight safety. Therefore, the present study aims to compare and contrast Taiwanese pilots' and foreign pilots' stressors, stress reactions, and stress coping behaviors.

Materials and Methods

Participants

From six airline companies in Taiwan, 727 civil pilots were invited to participate in this study. Most of them were male (99.0 percent), over 40 years of age (82.6 percent), having worked over ten years in their current companies (60.5 percent), married (84 percent), over 7000 flight hours (69 percent), and Taiwanese (82 percent).

Questionnaire

A researcher-created self-report questionnaire, using a 5-point scale, consisting of 104 forced-choice items related to national culture (22 items, the 5-point scale was from 1: no importance to 5: of utmost importance), stressors (25 items, the 5-point scale was from 1: not serious at all to 5: very serious), reactions under stress (27 items, the 5-point scale was from 1: disagree very much to 5: agree very much), self-rated job performances (12 items, 1: disagree very much to 5: agree very much), and stress coping behaviors (18 items, 1: never to 5: always), was used in this study.

The measures of the national culture were similar to the four types of national culture, namely individualism-collectivism, power distance (PD), uncertainty avoidance (UA), and masculinism-feminism as proposed by Hofstede (1980). The measures of stress coping behaviors were similar to Havlovic and Keenan's (1995) categories. The participants' demographic data and their self-reported experience of involving flight accidents or events were also collected.

Protocol and Data Analysis

The participants were invited to respond to the questionnaire individually while they were waiting for the physical examination. All of the collected data were processed and analyzed by employing SPSS 17.0.

Results

1. Local and Foreign Civil Pilots' National Culture-linked Characteristics

Since Factor Analysis and equamax rotation were used to group the measures of national culture as proposed by Hofstede (1980), the measures were clustered into four factors which, all together, accounted for 58.48 percent of the total variances. Based on the content meanings of each factor, the four factors were named as "masc-collectivism", "individualism", "masc-low PD", and "low UA", respectively. The factor "masc-collectivism" included the importance of personal growth, achievement, and work for country as well as organization. The factor "individualism" was composed of the items related to personal interests and benefits. The factor "masc-low PD" summarized the items involving personal achievement, challenge, adventure, and being consulted by direct superior. The "low UA" factor consisted of items related to less stress and tension at work, having considerable freedom to adopt personal methods to the job. Cronbach alphas, used to index the internal consistency of the items included in each factor, were .86, .82, .80, and .65, respectively.

As descriptive statistics showed that the civil pilots were likely to be "individualism" (mean = 4.07, SD = .60). Using t-tests to assess the score differences on these four factors in local and foreign pilots, foreign pilots were found to have higher scores on "low UA" ($t = -3.27$, $p < .001$) and greater "masc-low PD" scores ($t = -5.02$, $p < .001$) than Taiwanese pilots' scores (Figure 6.1).

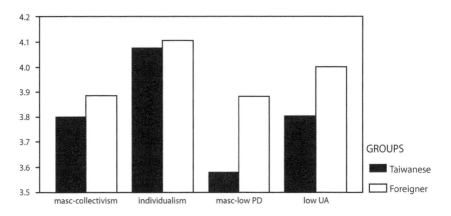

Figure 6.1   Comparisons between two groups of civil pilots' on cultural characteristics

2. Comparisons Between Local and Foreign Civil Pilots' Stressors

Factor analysis revealed that the 25 stressor items were clustered into four factors, accounting for 52.92 percent of the total variances. They were titled "interpersonal relationships", "leadership and management", "flight workloads", and "family and non-flight workloads." The factor "interpersonal relationships" depicted the items relating to relationships with co-workers, such as another pilot, crew members, ATCs' native accents. The factor "leadership and management" clustered the items of leadership of CP, rules and demands by company, difficulty in being promoted. The factor "flight workloads" grouped the items including not having enough rest between flights, too many flight hours, flight crews with conflicted relationships, and bad work climate. The factor "family and non-flight workloads" referred to lack of family support, having non-flight jobs, relationships with family, and burdensome training courses. Cronbach alphas for those four factors were .87, .80, .72, and .73, respectively. Descriptive statistics indicated that the top four ranks of stressors for the civil pilots were flight workloads (mean = 3.41, SD = .71), leadership and management (mean = 3.24, SD = .65), family and non-flight workload (mean = 2.99, SD = .76), and interpersonal relationships (mean = 2.82, SD = .58).

Moreover, independent-samples "t tests" indicated that the foreign pilots had higher scores on "flight workloads" as compared to those of Taiwanese (t = –3.87, p < .001). In contrast, Taiwanese pilots demonstrated more stress on "interpersonal relationships" (t = 6.02, p < .001) than foreign pilots (Figure 6.2).

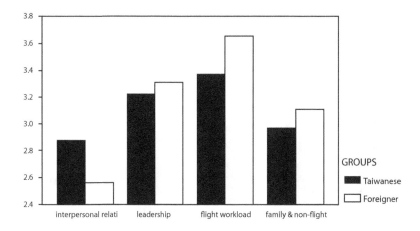

Figure 6.2   **Comparisons between two groups of civil pilots' stressors**

3. Comparisons Between the Two Groups of Civil Pilots' Stress Reactions

Factor analysis followed by Varimax rotation indicated a four-factor structure that can be used to explain 55.5 percent of the total variances. According to each factor's items, these factors were titled emotional reactions (e.g., anxiety, depression, irritation), behavioral reactions (e.g., slower reactions), physical reactions (e.g., sleep problems, headache), and poor concentration (e.g., need to re-read or pay extra attention to understand), accordingly. Their Cronbach alphas were .92, .86, .81, and .50, respectively.

Independent-samples "t tests" suggested that local and foreign pilots displayed different reactions on all of these measures while experiencing a stressful situation (all p values were less than .001) (Figure 6.3). Obviously, Taiwanese pilots tended to have stronger reactions than foreign pilots. However, both Taiwanese and foreign pilots demonstrated seriously poor concentration when they were in a stressful condition (mean = 3.50, SD = .68).

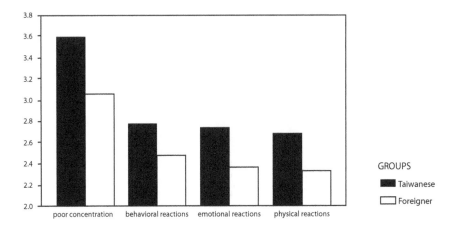

Figure 6.3   Comparisons between Taiwanese and foreign civil pilots' stress reactions

4. Taiwanese and Foreign Civil Pilots' Habitual Coping Behaviors

As factor Analysis and Varimax rotation were used to examine the validity for coping behaviors, five factors were determined. The five factors together can explain 65.6 percent of the total variances. Based on the content analysis of each factor, five factors were named after the same titles used by Havlovic and Keenan (1995), they were direct actions, positive thinking, seeking help, avoiding stress, and using alcohol/tobacco. As the reliability analysis was done, these five factors' Cronbach alphas ranged from .88 to .68.

Descriptive statistics showed the civil pilots participating in this study preferred to use direct actions (mean = 3.63, SD = .54) and positive thinking (mean = 3.54, SD = .63) in response to their stressful situations. Independent-samples "t tests" further showed that foreign pilots were more prone to generate positive thinking to cope with their stress than Taiwanese pilots (t = –2.02, p < .05). Interestingly, foreign pilots were more prone to request help (p = .056), and less prone to use alcohol or tobacco (p = .057) to cope with their stress than did Taiwanese pilots. These results were shown in Figure 6.4.

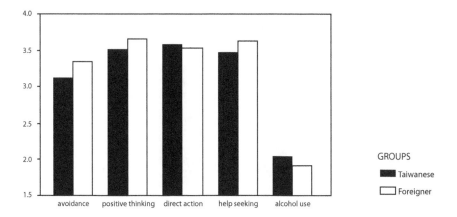

Figure 6.4   Taiwanese and foreign civil pilots' habitual coping behaviors in response to stressful situations

5. Comparisons Between the Two Groups of Civil Pilots' Perceptions for Their Work Performance

Factor analysis and Varimax rotation results revealed that twelve items related to job performance could be clustered into three factors, which together accounted for 65.56 percent of the total variances. These three factors were called achievement at work, job ability, and good peer relationship, and their Cronbach alphas were .90, .71, and .63, accordingly.

Moreover, descriptive statistics demonstrated that these civil pilots had the highest mean scores on "job ability" (mean = 4.05, SD = .54), and the lowest mean scores on "achievement at work" (mean = 3.21, SD =.76). Independent-samples "t tests" showed that Taiwanese and foreign pilots both believed that they were very professional (job ability) and had very good relationships with their co-workers (good peer relationship). However, it was of interest to note that Taiwanese pilots even felt better about achievement at work (t = 6.50, p < .001) and good peer relationship (t = 3.66, p < .001) than did foreign pilots. These results were shown in Figure 6.5.

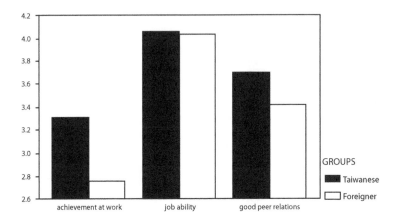

Figure 6.5   Comparisons between the two groups of pilots' current work situations

6. Correlations Among Stressors, Stress Reactions, Coping Behaviors, and Work Performance

The correlations among the variables of stressors, stress reactions, coping behaviors, and work performance were obtained according to Pearson correlation analyses (see Table 6.1).

In Table 6.1, all factors within stressors, stress reactions, and performances were correlated in a positive manner (all p values < .001). In the coping behaviors category, "using alcohol/tobacco" was positively correlated with "avoiding stress" (r = .15, p < .001), but negatively correlated with "direct actions" (r = −.08, p < .05). It was of interest to note that pilots with a higher "interpersonal relationships" stressor were more prone to choose "avoiding stress" in coping behaviors (r = .18, p < .001) and using alcohol/tobacco (r = .16, p < .001). The civil pilots with greater stressors regarding "leadership and management" were more likely to use all five types of coping behaviors (all p values were less than .05). Pilots with a source of stressors, including "flight workloads", "family and non-flight workloads" were prone to adopt all possible coping behaviors except "using alcohol/tobacco". In addition, the "leadership and management" stressor was negatively correlated with "achievement at work" (r = −.22, p < .001) and "good peer relationships" (r = −.07, p < .05); while "flight workloads" stressor was negatively correlated with "achievement at work" (r = −.17, p < .001).

Interestingly and surprisingly, it was demonstrated for the first time that pilots prone to use the control orientated adaptive coping behaviors such as "positive thinking", "direct actions", and "seeking help", were found to have the very best self-evaluated performances. In contrast, pilots prone to habitually take "avoiding stress" coping behavior evaluated themselves as having a low degree of *p<.05; **p<.01; ***p<.001

Table 6.1   Correlations among stressors, stress reactions, coping types, and performance

	1	2	3	4	5	6	7	8	9	10	11	12	13	14	15	16
Stressors																
1. interpersonal relationship	1															
2. leadership	.56***	1														
3. flight workloads	.46***	.52***	1													
4. family & non-flight	.42***	.58***	.50***	1												
Stress reactions																
5. poor concentration	.26***	.15***	.13***	.10*	1											
6. behavioral reactions	.36***	.23***	.24***	.22***	.34***	1										
7. emotional reactions	.32***	.23***	.23***	.19***	.32***	.74***	1									
8. physical reactions	.34***	.21***	.18***	.15***	.27***	.61***	.67***	1								
Coping behaviors																
9. avoid	.18***	.20***	.24***	.16***	.21***	.31***	.34***	.27***	1							
10. positive thinking	.00	.10*	.14***	.17***	.06	.04	.07	-.09*	.22***	1						
11. direct action	.05	.13***	.14***	.14***	.14	.04	.05	-.03	.31***	.65***	1					
12. seeking help	.04	.10*	.12**	.13***	.11**	.06	-.00	-.01	.18***	.47***	.47***	1				
13. using alcohol	.16***	.10*	.03	.02	.03	.22***	.22***	.21***	.15***	-.06	-.08*	.01	1			
Performances																
14. achievement at work	.03	-.22***	-.17***	-.07	.11***	.11***	.06***	.08***	-.14***	.09*	.12**	.20***	.03	1		
15. job ability	-.05	-.05	-.01	-.05	.12**	-.01	-.03	-.04	.05	.22***	.24***	.19***	-.10**	.29***	1	
16. good peer relationship	.01	-.07*	-.03	-.03	.12**	.04	-.01	-.03	-.02	.16***	.21***	.21***	-.02	.45***	.38***	1

achievement at work (r = −.14, p < .001). Last but not least, pilots prone to "using alcohol/tobacco" were self-rated as having low "job ability" (r = −.10, p < .01).

Discussion

In this empirical study, it was established that Taiwanese and foreign civil pilots working for local airline companies indeed had different stressor profiles and habitual stress coping behaviors. Specifically, Taiwanese pilots appeared to exhibit stronger stress reactions than foreign pilots. Nonetheless, a consistent self-disclosure from both Taiwanese and foreign pilots was that they were very concerned about poor concentration when they were in a stressed condition. Since poor concentration might render decreased ability of cognitive processing and jeopardize the flight safety, the implication of these results was that an early-stage, reliable, and valid detection system for pilots' stress reactions should be developed in the future.

As for multi-cultured airline companies, it is suggested much attention and effort be paid to matching pilots' perceptions regarding their work situations (especially operation rules, work climate, and flight performance-based promotion) with their personal needs. Obviously, civil pilots raised and trained in various cultures displayed, to some degree, different expectations and perceptions in work situations and personal needs. In order to reduce human errors in the cockpit and enhance flight safety, culture-based matching of pilots' attitudes, expectations and actual work situations could be a very promising policy for those multi-cultured airline companies.

Likewise, the findings that Taiwanese and foreign pilots exhibit different stressor profiles and coping behaviors suggest that it is necessary to design suitable intervention training programs to encourage and facilitate culture-based stress coping behaviors. For those pilots who are accustomed to adopting "avoiding stress" and "using alcohol/tobacco" coping behaviors, it seems to be imperative to develop an early-staged monitoring system for these ineffective coping mechanisms. An effective training program should be developed accordingly to encourage pilots' advantageous coping behaviors. As a matter of fact, understanding and respecting pilots' original cultures can be win-all condition for enhancing pilots' achievement, performances and flight safety.

Acknowledgments

This study was, in part, supported by a NSC grant (No. 96-2221-E-309-017) to Dr. C.G. Cherng. The authors thank Dr. L. Yu for providing constructive suggestions.

References

Aldwin, C.M. and Revenson, T.A. (1987). Does coping help? A re-examination of the relation between coping and mental health. *Journal of Personality and Social Psychology*, 53, 337–48.

Alkov, R.A., Gaynor, J.A. and Borowsky, M.S. (1985). Pilot error as a symptom of inadequate stress coping. *Aviation, Space, and Environmental Medicine*, 56, 244–7.

Auerbach, S.M. and Gramling, S.E. (1998). *Stress Management Psychological Foundations*. New Jersey: Prentice Hall.

CAA (2010). Civil air transportation statistics. Retrieved September 6, 2010 from <http://www.caa.gov.tw>.

Cooper, C.L., Cooper, R.D. and Eaker, L.H. (1988). *Living with Stress.* Penguin Books.

DuBrin, A.J. (2004). *In Applying Psychology: Individual and Organizational Effectiveness*. New Jersey: Pearson Education.

Green, R.G., Muir, H., James, M., Gradwell, D. and Green, R.L. (1996). *Human Factors for Pilots*. Aldershot: Ashgate.

Havlovic, S.J. and Keenan, J.P. (1995). Coping with work stress: the influence of individual differences. In: R. Crandall and P.L. Perrewe (eds), *Occupational Stress: A Handbook*, 179–92. USA: Taylor & Francis.

Helmreich, R.L. and Merritt, A.C. (1998). *Culture at Work in Aviation and Medicine: National, Organizational, and Professional Influences*. Aldershot: Ashgate.

Hofstede, G. (1980). *Culture's Consequences: International Differences in Work-related Values*. Beverly Hills, CA: Sage.

Hofstede, G. (1991). *Cultures and Organizations: Software of the Mind*. Maidenhead: McGraw-Hill.

Holmes, T.H. and Rahe, R.H. (1967). The social readjustment rating scale. *Journal of Psychosomatic Research*, 11, 213–18.

Hui, C.H. and Triandis, H.C. (1986). Individualism-collectivism: a study of cross-cultural researchers. *Journal of Cross-Cultural Psychology*, 17, 222–48.

Kushnir, T. (1995). Stress in ground support personnel. In: J. Ribak, R.B. Rayman and P. Froom (eds), *Occupational Health in Aviation: Maintenance and Support Personnel*, 51–72. New York: Academic Press.

Li, G., Baker, S.P., Grabowski, J.G. and Rebok, G.W. (2001). Factors associated with pilot error in aviation crashed. *Aviation, Space, and Environmental Medicine*, 72, 52–8.

Selye, H. (1974). *Stress Without Distress*. Philadelphia, PA: Lippincott.

Spence, J.T. (1985). Achievement American style: the rewards and costs of individualism. *American Psychologist*, 40, 1285–95.

Thom, T. (1997). *Human Factors and Pilot Performance: Safety, First Aid and Survival*. UK: Airlife.

Zuckerman, M. (1991). One person's stress is another person's pleasure. In: C.D. Spielberger, I.G. Sarason, Z. Kulcsar and G.L. Van Heck (eds), *Stress and Emotion: Anxiety, Anger, and Curiosity*, Vol. 14. New York: Hemisphere Publishing Corporation.

Chapter 7

Anticipatory Processes in Critical Flight Situations

K. Wolfgang Kallus

Karl-Franzens Universität Graz, Austria

Introduction

In the following section, arguments and empirical evidence are presented that elucidate the prominent role of anticipatory processes in piloting an aircraft – especially in critical flight situations.

Anticipatory processes are needed in a scientific description of behavior in highly dynamic environments. Part of this behavior can be explained by habits that are activated quickly enough to account for the changes in the environment and are far too quick to be accounted for in a mode of reacting to that which has been perceived in the situation. In action theoretical conceptions, we have highly automated processes that do not need conscious and attention demanding processing and reactions. Moving with high velocity, as pilots do when flying an aircraft, is a most demanding task for our information processing system. Without a proper way to anticipate the upcoming environment and events, many of our motor activities would be too slow and our actions would easily fall behind the stream of events. Even simple organisms like flies have to account for dynamics when flying – and their eye movements do. The principle of reafference has been developed by the Nobel-prize winners von Holst and Mittelstaedt (1950) to account for this fact and the fact that it is necessary to have a process that allows us to decide that a movement has been successful in a dynamic environment. Hoffmann's model of Anticipatory Behaviour Control is an action theoretic conception that stresses the role of cognitive "top-down processes". Cognitive "top-down processes" have been a matter of debate in scientific psychology with a long tradition. Since the time of William James and his ideomotor principle (1890–1981), action oriented concepts have been proposed in order to account for top-down processes of behavior. Important and more recent contributions to the role of anticipation in psychomotor behavior and perception stem from Neisser (1976), Prinz (1983) and Hommel, Musseler, Aschersleben, and Prinz (2001). Interestingly enough, these approaches have always been a kind of theoretical counterpart to the behavioristic S-R-Models or SORKC Models (Kanfer and Phillips, 1979). Anticipations are cognitive events that have not been observable directly with sufficient precision for a long period of research. Recent developments to combine behavioral neuroscience approaches, psychophysiological approaches, reconstruction techniques and behavioral observations in well defined paradigms make it possible to overcome the limitations of the old "introspective methodologies". These were the basic methods of the early cognitive "research schools" like the "Würzburg School" (Külpe, 1922) who dealt with early developments of the relation between motivational and anticipatory components of behavioral organization (Ach, 1910). A more recent approach to integrate learning theory with anticipatory processes has been elaborated by Hoffmann (1993, 2003). His model explains why rewards work. By letting the individual expect the desired effect, rewards enhance the probability of a rewarded behavior in a given situation. Thus anticipations are not only functional in a dynamic environment by anticipation of future changes in the environment, anticipation might also be relevant with respect to its content.

"I work on the basis of my mental traffic picture" is a standard answer one gets, when beginning a task analysis in air traffic control. This can be viewed as a concrete formulation of the idea of Hoffmann's principle of Anticipatory Behaviour Control (2003). Hoffmann, and Kunde, Elsner and Kiesel (2007) have reformulated the classical operant conducting approach with an "anticipation" oriented perspective. Controllers "predict" or "anticipate" how the situation will develop with (and without) their interventions based on their traffic picture. Anticipatory Behaviour Control (ABC) can be viewed as a basic process that allows an operator to act "proactively" in a dynamic system. Thus, ABC should be viewed as a primary and basic process for pilots in critical fight situations in which "reacting" might be far too slow. The ABC concept is well compatible with Endsley's (1995, 2002) well known concept of situational awareness – but it changes the focus drastically, as anticipation of future events is viewed as the first and basic process, which does not necessarily imply an "understanding" of the situation (Kallus, 2009). Anticipatory processes are organized on different levels of information processing and interact to establish situation awareness and states like spatial orientation. A basic model of the role of anticipatory processes in situational awareness (Kallus, 2009) is depicted in Figure 7.1.

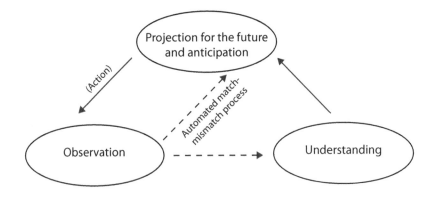

Figure 7.1   ABC in pilots

The model can equally be used to describe the state of spatial orientation. Spatial orientation can be defined as the ability "... to sense correctly the position, motion or attitude of his/her aircraft or of him/herself within the fixed coordinate system provided by the surface of the earth and the gravitational vertical" (Benson, 1999, p. 419). The continuous anticipation of the forthcoming situation and the continuous matching with the information from the proximal and distal environment and from the relevant displays is a basic process to maintain situational awareness and to keep the pilot in proper spatial orientation. In addition to the anticipation – action – comparison process, active checking of the relevant information (especially on the displays) in discontinuous intervals seems to be a characteristic of operators in dynamic systems (Kallus, Barbarino and Van Damme, 1998). The checking of instruments in a "T-system of gaze" is part of the basic training of pilots. The strongly automized processes of anticipation – action – comparison loops has to be supplemented by active checking processes to establish situation awareness. The active checking is supported by checklists in many aviation tasks. Thus, automatized as well as explicit anticipations govern the activity of operating and managing a dynamic system such as an aircraft or a sector of airspace. This concept was first developed on the basis of Europe-wide interviews conducted within a cognitive task analysis of air traffic controllers (Kallus, Dittmann, Barbarino and Van Damme, 1999). In this task, analysis controllers stated uniquely that they base their action on a (cognitive) traffic picture. This is in full accordance with other descriptions of the controllers' work (Hopkin, 1995; Stein, 1998). The dynamic traffic picture allows the controllers to project the traffic into the future. Based on this projection, they provide efficient and safe separation and guidance of the air traffic (Kallus, Barbarino, and Van Damme, 1998). The role of anticipatory processes is central in air traffic control, but it is even more important in the actual flight situation of pilots. In a "reactive" mode of action, pilots would always fall "behind the actual situation" and risk spatial disorientation.

Spatial orientation of pilots relies heavily on a continuous anticipation process. Spatial disorientation can be viewed as a faulty or disturbed anticipation – action – comparison process in most instances due to perceptual illusions.

A couple of results from our studies with disorientation prone situations in the simulator stress the assumption that Anticipatory Behaviour Control plays a central role in coping with difficult flight situations.

One of the most striking results was obtained by Kallus and Tropper (2004) with military pilots who had to manage a landing at night with a faintly illuminated runway. This situation is called a "black-hole-approach" and is accompanied by an underestimation of the angle of the glide path during landing. Thus pilots tend to fly too low and crash during landing if they do not correct the misperception. Twenty-four military pilots with valid flight licenses were studied, using the DISO Airfox Simulator of AMST Systemtechnik GmbH in Ranshofen, Austria with an F18 jet model. Pilots were subdivided ex post into three performance groups: correct landing, managing problems adequately and crash. The exercise was part of an experiment in which pilots were trained to better cope with disorientation prone flight situations by correcting their perceptual illusion using unobtrusive information from the flight instruments. During this experiment, performance was assessed along with psychophysiological recordings and subjective state. Figure 7.2 shows the baseline corrected heart rate of the three performance groups during the black hole approach.

The three groups show comparable heart rates before they receive control from the instructor pilot and they show a comparable initial rise when they pass the 14-miles-out trigger. The group, which is about to crash later, develops a marked rise in heart rate with peaks 60 and 30 seconds before they crash. Pilots with an adequate landing also show a continuous rise in heart rate, but with a much lower rate of increase. Pilots who manage the situation safely but with a touch and go or go-round manoeuvre show an intermediate increase with a peak about 40 seconds before the expected landing. The crash group shows an increase in heart rate, which indicates a different anticipatory process. Instead of recognizing the problem and executing a safe solution as done in the group "problems", they continue their path to crash. Their anticipatory state shows a marked over-activation, but their behavior was not affected correspondingly. The result seems to indicate that anticipatory processes do not necessarily cross the border of awareness. The result of changed anticipatory heart rates for pilots who do not manage the black hole approach adequately could be replicated with VFR pilots in the master thesis of Zauner (2006, cf. Kallus, 2009).

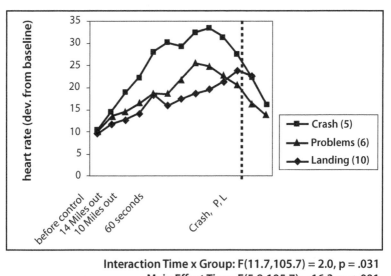

Interaction Time x Group: $F(11.7, 105.7) = 2.0$, $p = .031$
Main Effect Time: $F(5.9, 105.7) = 16.3$, $p < .001$

Figure 7.2   HR black hole approach

Critical flight situations like unusual attitude recoveries, flight into IMC for VFR pilots and spatial disorientation phenomena due to perceptual illusions are most taxing situations, especially for military pilots and VFR pilots. Spatial disorientation is a major source of fatal accidents in both areas (VFR and military). In the military area, both jets and helicopters have to be considered for training or support procedures. Two ways to assist pilots in critical flight situations have been researched recently. The "American" approach advocates assisting pilots with warnings and technical tools in disorientation prone situations. Computer systems have been developed to indicate disorientation prone situations and critical attitudes of the aircraft. The information is integrated into new cockpit displays or verbal information. This system helps pilots to recover with fewer errors form unusual attitudes, i.e., coping with disorientation prone situations in a more appropriate way. This could at least be shown in a low fidelity simulation (Wickens, Self, Andre, Reynolds, and Small, 2007; Wickens, Small, Andre, Bagnall, and Brenaman, 2008).

The "Austrian" approach follows a more human-centered direction and developed a training system that provides pilots with the skills to anticipate and cope with critical flight situations. This allows the pilots to act and decide proactively in critical flight situations. The result cited above stems from a training study in which military pilots were repeatedly confronted with typical disorientation situations in the simulator and successfully trained to cope with them – mostly by relying primarily on instrument information (Kallus and Tropper, 2004).

Another training study revealed further evidence on the central role of correct anticipatory processes in managing critical flight situations (Kallus, Tropper, and Boucsein, 2011). VFR pilots were trained to manage critical flight situations either with the motion function of the simulator switch on or with the motion function switched off. Both groups were compared to a free flight control group that did not obtain specific training. Figure 7.3 shows the results for the most difficult "unusual attitude recovery" of the test session after training. While the subjects of the training group with motion on recovers in about 25 seconds, the training group with motion switched off during training needs more than 40 seconds on average. The performance is worse than the control group. Analysis of variance showed a highly significant group effect. Post hoc Tukey tests showed a remarkable difference between the two training groups. Similar effects were obtained in other exercises (Kallus, Tropper, and Boucsein, 2011).

The results indicate that without motion no adequate anticipation had been established.

This approach leads to a reformulation of current situational awareness models. Endsley's sequence of perceiving, understanding and projection into the future is turned upside down if the projection into the future becomes the first and leading step. In addition, we would include a loop of non-conscious, highly automated processes in a model of "Anticipatory Behaviour Control" as depicted in Figure 7.1.

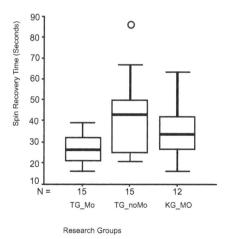

Figure 7.3   Spin recovery

Finally, a training study with VFR pilots was conducted in which the training elements were explicitly designed to optimize anticipatory processes during critical flight manoeuvres (Koglbauer, 2009; Boucsein, Koglbauer, Braunstingl, and Kallus, 2011). The exercises were shaped similarly to mental training conceptions, using brief mental action oriented code words (e.g., "pull") to mark the critical steps of the manoeuvre. The study with 24 licenced VFR pilots (no IFR licences, no acrobatic experience, less than 450 hours flight experience) involved a pre-training demonstration flight in an acrobatic aircraft with an acrobatic aircraft training instructor and a check flight in the simulator and two check flights in the aircraft after training. Psychophysiological recordings of heart rate and electrodermal activity were obtained to reflect anticipatory processes during the check flight in the simulator and in the aircraft. Subjects of the control group had the pre-training demonstration flight and both check flights and an equal time of flying the simulator. They had to conduct complex navigation tasks in the simulator instead of the training of critical flight situations.

Figure 7.4 shows, that the two groups had quite similar emotional activation in the (passive) initial flight in the acrobatic bi-plane. Values for the experimental group drop from initial flight to the test flight in the simulator, while values for the control group increase when they have to manage the exercises themselves. Both groups show an increase from simulator to active test flights in the aircraft; however the training group keeps the emotional activation on a considerably lower level. The results in Figure 7.4 indicate that the training was able to change the activation processes during the task execution. This can be shown in the simulator as well as in the aircraft. This result is important theoretically with respect to stress levels as well as for the role of anticipatory processes in critical fight situations.

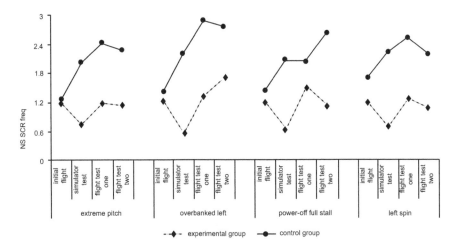

Figure 7.4   EDA

In a more detailed analysis the exercise was subdivided in action related phases: anticipation, execution, and post execution (recovery). Results for the anticipatory period of each manoeuvre are shown in Figure 7.5.

Figure 7.5 indicates that differences in activation between the training group and the control group can already be shown clearly in the anticipatory phase. This means that the anticipation directed training was able to affect the psychological processes before (!) the task was executed. It shows that the training was not just affecting motor responses.

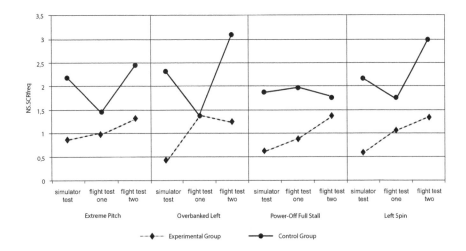

Figure 7.5   EDA anticipation

From a practical point of view, data indicate that the training is able to reduce anticipatory emotional tension. This seems to be a most important result as we find a panic-like emotional state in VFR pilots in many instances preceding fatal accidents. One striking event in Austria appeared in 2007 when a German VFR pilot and his co-pilot/passenger crashed into the mountains near lake Ossiach. They seemed to have changed direction about 180 degrees opposite to the airport after their last contact to the tower in bad weather (BFU, 2007) instead of approaching the airport.

All in all, this chapter tried to promote some arguments to account for anticipatory processes in flying in an explicit and systematic way. This is intended to add additional aspects of anticipatory processes in accident-incident analysis, in training, in monitoring critical psychological states of pilots, in the analysis of situation awareness and in selection. The arguments should not be misread as arguments to stop the development of technical aids in disorientation-prone environments. The proper combination of different approaches will help to reduce the number of fatal accidents such as controlled flight into terrain due to spatial disorientation and/or panic-like behavior in the cockpit. From a conceptual point of view, the automation oriented approach seems viable to better deal with everyday situations due to lapses in attention and easily predictable sensory illusions (like the pitch up illusion), while the human centered approach allows pilots to improve performance in unexpected critical situations and to cope better with rare critical events.

References

Ach, N. (1910). *Über den Willensakt und das Temperament.* Leipzig: Quelle and Meyer.

Benson, A.J. (1999). Spatial disorientation – general aspects. In: J. Ernsting, A. Nicholson and D.J. Rainford (eds), *Aviation Medicine.* Oxford: Butterworth Heinemann.

BFU (2007). *Jahresbericht 2007.* Braunschweig: Bundesstelle für Flugunfalluntersuchungen.

Boucsein, W., Koglbauer, I., Braunstingl, R., and Kallus, K.W. (2011). The use of psychophysiological measures during complex flight maneuvers – an expert pilot study. In: M. Ouwerkerk, J.H.D.M. Westerink and M. Krans (eds), *Sensing Emotions in Context. The Impact of Context on Behavioural and Physiological Experience Measurements.* (pp. 44–64). Berlin: Springer.

Endsley, M.R. (1995). Toward a theory of situation awareness in dynamic Systems. *Human Factors*, 37(1), 32–64.

Endsley, M.R. (2000). Theoretical underpinnings of situation awareness: a critical review. In: M.R. Endsley and D.J. Garland (eds), *Situation Awareness Analysis and Measurement.* (pp. 3–28). Mawah, NY: Lawrence Erlbaum.

Hoffmann, J. (1993). *Vorhersage und Erkenntnis: Die Funktion von Antizipationen in der menschlichen Verhaltenssteuerung und Wahrnehmung.* [*Anticipation and Cognition: The Function of Anticipations in Human Behavioral Control and Perception*]. Göttingen: Hogrefe.

Hoffmann, J. (2003). Anticipatory behavioral control. In: M. Butz, O. Sigaud, and P. Gerard (eds), *Anticipatory Behavior in Adaptive Learning Systems.* (pp. 44–65). Heidelberg: Springer.

Hommel, B., Musseler, J., Aschersleben, G., and Prinz, W. (2001). The theory of event coding (TEC): a framework for perception and action planning. *Behavioral and Brain Sciences*, 24, 849–78.

Hopkin, V.D. (1995). *Human Factors in Air Traffic Control.* London: Taylor and Francis.

James, W. (1981). *The Principles of Psychology (Orig. 1890).* Cambridge, MA: Harvard University Press.

Kallus, K.W. (2009). Situationsbewusstsein und antizipative prozesse. [Situation awareness and anticipatory processes] *Zeitschrift für Arbeitswissenschaft.* 1, 17–22.

Kallus, K.W., Barbarino, M., and Van Damme, D. (1998). *Integrated Task and Job Analysis of Air Traffic Controllers Phase 1: Development of Methods* (HUM. ET1.STO1.1000-REP-03). Brussels: Eurocontrol.

Kallus, K.W., Dittmann, A., Barbarino, M., and Van Damme, D. (1999). Basic cognitive processes of air traffic controllers. In: D. Harris (ed.), *Engineering Psychology and Cognitive Ergonomics Volume Three.* (pp. 113–120). Aldershot: Ashgate.

Kallus, K.W., and Tropper, K. (2004). Evaluation of a spatial disorientation simulator training for jet pilots. *International Journal of Applied Aviation Studies*, 4(1), 45–55.

Kallus, K.W., Tropper, K. and Boucsein, W. (2011). The importance of motion cues in spatial disorientation training for VFR pilots. *International Journal of Aviation Psychology*, 21(2), 135–52.

Kanfer, F.H. and Phillips, J.S. (1979). *Learning Foundations of Behavior Therapy.* New York: Wiley and Sons.

Koglbauer, I. (2009). Multidimensional approach of threat and error management training for VFR pilots. Unpublished dissertation. Karl-Franzens University of Graz.

Külpe, O. (1922). *Vorlesungen über Psychologie (2.Aufl.).* Leipzig: S. Hirzel.

Kunde, W., Elsner, K. and Kiesel, A. (2007). No anticipation–no action: the role of anticipation in action and perception. *Cognitive Processing*, 8(2), 71–8.

Neisser, U. (1976). *Cognition and Reality.* San Francisco: Freeman.

Prinz, W. (1983). *Wahrnehmung und Tätigkeitssteuerung.* Heidelberg: Springer.

Stein, E.S. (1998). Human operator workload in air traffic control. In: M.W. Smolensky and E.S. Stein (eds), *Human Factors in Air Traffic Control.* (pp. 155–84). London: Academic Press.

von Holst, E. and Mittelstaedt, H. (1950). Das Reafferenzprinzip. [The reafference principle] *Naturwissenschaften*, 37, 464–76.

Wickens, C.D., Self, B.P., Andre, T.S., Reynolds, T.J., and Small, R.L. (2007). Unusual attitude recoveries with a spatial disorientation icon. *International Journal of Aviation Psychology*, 17(2), 153–65.

Wickens, C.D., Small, R.L., Andre, T., Bagnall, T., and Brenaman, C. (2008). Multisensory enhancement of command displays for unusual attitude recovery. *International Journal of Aviation Psychology*, 18, 255–67.

Zauner, Ch. (2006). Experimentelle Untersuchung psychophysiologischer Reaktionen auf verschiedene Flugprofile. [Experimental study of psychophysiological reactions caused by different flight profiles]. Unpublished thesis. Karl-Franzens-University of Graz.

Chapter 8

Error Detection During Normal Flight Operations: Resilient Systems in Practice

Matthew J.W. Thomas and Renee M. Petrilli
University of South Australia, Australia

Introduction

Human error continues to be implicated as a causal factor in accidents in high-risk industries worldwide. As Wreathall and Reason (1992) have so eloquently stated, the history of accidents is also the history of the human contribution to accidents. While human error is frequently seen as indicative of poor performance or aberrant behavior, this view is both counter-productive and typically untrue from the perspective of aviation safety. It is now well established that error is a natural part of human performance, and frequently unavoidable in day-to-day work activities. To this end, understanding the predictable aspects of human error in high-risk work environments, and examining the relationship between error occurrence and error management, form important new frontiers for aviation safety.

Accepting the Ubiquity of Human Error

It is now several decades since the science of Human Factors introduced the concept of the ubiquity of human error to high-risk industries such as aviation. Seminal texts such as James Reason's *Human Error* (1990) translated considerable psychological research into a practical manifesto for new perspectives on safety management. Much of this new focus on human error was placed on expanding our understanding of the aetiology of human error. Specific points of investigation included identifying the range of error producing conditions, as well as the influence of factors such as workload, and of system and interface design on the generation of error.

Human Error was frequently conceptualized as the end of a trajectory, the culmination of pre-existing systemic conditions that ultimately contributed to the occurrence of error (Reason, 1997). This approach gave rise to a range of models that attempted to enable the classification of errors and their precursors, such as the Human Factors Analysis and Classification System – "HFACS" (c.f. Shappell and Wiegmann, 2000, 2001; Wiegmann and Shappell, 2001). Within this framework, emphasis was still applied to the minimization of the occurrence of error through addressing the "upstream" factors that precipitated error. Error management was seen in terms of better system design and the removal of error producing conditions.

Considerable existing research examining the general occurrence of Human Error across a wide range of everyday and work environments has informed systems for the classification of error. Through the classification of error, we risk creating an illusion of understanding the causal factors involved through a simplistic process of relabelling and grouping similar types of error. However, it is possible to build effective mechanisms for safety-related change through analyses of error that de-emphasize the construction of cause and focus on the identification of patterns in error occurrence. Understanding the "genotypical mechanisms of failure" elucidate the means by which operators create safety in practice, and map universal patterns of safety breakdown (Dekker, 2003).

There is no doubt that this approach to error management resulted in significant enhancements to safety. However, this approach failed to emphasize the inevitability of error and its spontaneous occurrence even within the best-designed systems. The next step was developing an approach that accepted both the ubiquity and inevitability of human error and therefore focussed not only on error reduction, but also on the subsequent management of error to either mitigate or ameliorate any effects of error on system performance.

The Rise and Rise of Threat and Error Management

Within the context of error mitigation and amelioration, perhaps the most popular of the recent models is that of Threat and Error Management (TEM). Developed predominantly from Professor Robert Helmreich's laboratory at the University of Texas, the TEM model emphasizes the positive functions involved in the *recovery* from inevitable error (Helmreich, 2000; Helmreich, Merritt, and Wilhelm, 1999). The TEM model offers a refreshingly honest appraisal of system safety, and sees error as a natural part of everyday expert performance. The model emphasizes the recovery processes inherent in expert performance and the ways in which error is *detected* and *managed*.

More recently, resilience engineering has focussed on error as natural performance variability. Resilience engineering describes safety in terms of helping people cope with the complexity of complex socio-technical systems of work. Rather than attempting to tabulate error in order to design interventions to reduce this error, resilience engineering places emphasis on the adaptive functions that enable tolerance of normal human performance variability (Hollnagel, Woods, and Levenson, 2006). Like resilience engineering, TEM has also shared some of this philosophy, yet, perhaps in a more operationally meaningful way. The explicit focus of these new approaches to human error is simply to anticipate that error will occur, and that a safe system is one in which error can be detected and mitigated prior to any potential impacts on system integrity.

Importantly, in most high-risk industries such as commercial aviation, the skilled expert operator is typically afforded the opportunity to detect, manage, and recover from the errors that occur as a natural part of expert performance. Recent research has highlighted the fact that a significant proportion of errors remain undetected during normal flight operations (Thomas, 2004). However, relatively little is known about what types of errors remain undetected, and what are the primary mechanisms for error detection.

One step towards a better understanding of the patterns of error occurrence in normal flight operations involves the analysis of the interaction between error type, and the detection of that error. Not all errors are a result of the same cognitive processes, nor can be described in terms of the same causal mechanisms. Similarly, a wide range of mechanisms can be shown to enable the detection and effective management of errors in the complex environment of normal flight operations. It is possible that certain types of errors are more likely to be detected, and that the underlying genotype may play an important role in the detection and effective management of error.

This chapter presents an exploratory study of the interaction between error type and error detection mechanism in the context of normal flight operations within the commercial airline environment.

Method

Participants

Data were collected from a total of 215 sectors of normal line operations in a single-aisle jet fleet of a large commercial airline. All participants volunteered to have a jump-seat observer collect data during the flight operation.

Design and Procedure

The study adopted a naturalistic observational design, and utilized a structured performance analysis methodology for the collection of detailed data relating to error occurrence and management. The performance analysis methodology was primarily based on a model of Threat and Error Management developed by the University of Texas, and as utilized in the Line Operations Safety Audit (LOSA) methodology (Helmreich, Klinect, and Wilhelm, 1999; Klinect, 2002; Klinect, Wilhelm, and Helmreich, 1999).

A team of seven expert observers, who received formal training in the observation methodology, collected data. The observer training process involved three major components: 1) two days of classroom training and initial observer calibration; 2) a day of practice observations during normal line operations; and 3) a final day of feedback on practice observations and observer calibration. Purpose-built video examples of crews from the airline, flying segments of simulated line operations, were utilized as discreet case studies for observer training and calibration. As an innovative new element of observer training, these specially created videos facilitated observer calibration through the practice analysis of naturalistic crew performance, as well as providing invaluable exposure to the Standard Operating Procedures (SOPs) and operational context of the airline involved in the study.

Data were collected from a total of 215 sectors of normal line operations. The data collected during these flights were subjected to a formal data cleaning process to ensure validity. Data cleaning involved the analysis of each error event by members, including representatives of the airline's safety, training, standards and operations departments. This process ensured accuracy in the data-set and that the observers interpretation of erroneous crew actions were accurate, especially with reference to the SOPs of the airline involved in the study.

Subsequent to data cleaning, further post hoc analysis of the data was undertaken involving the coding of error events according to the taxonomies of error phenotype and genotype described below.

Measures

Errors were coded under the three traditional categories of error type, namely: 1) slips; 2) lapses; and 3) mistakes (Reason, 1987, 1990). This form of categorization provided an indication of whether an error had occurred in the planning or execution stages of crew action.

Furthermore, each error was subjected to specific coding of the mechanisms by which errors were detected by the observed crews. Eight categories of error detection mechanism were utilized within this coding framework, namely: 1) Aircraft Warning System; 2) Aircraft Status or Performance; 3) Cross Check (Monitoring Other); 4) Scan (Monitoring Self); 5) Checklist; 6) External Person; 7) None (Not Detected) and 8) Undeterminable.

Analyses

Data were subjected to descriptive analyses, and the Chi Squared statistic was utilized to test for significant differences in the distribution of error detection mechanism as a function of error genotype.

Results

From the 215 sectors of normal flight operations observed in this study, a total of 695 individual error events were observed and subjected to coding and analysis.

With respect to error type, lapses were most frequently observed in this data-set, indicating a prevalence of memory-related errors occurring during the execution of planned actions. The overall distribution of error type is provided in Table 8.1, below.

Table 8.1   Distribution of error type

Generic Error Process	
Slip	20.3
Lapse	48.5
Mistake	31.2

Frequency expressed as a percentage of all errors (N = 695)

With respect to error detection mechanism, the largest proportion of errors were detected through cross-checking, where a crew-member detected an error which was made by another crew-member. Of the errors, more than one third remained undetected. The overall distribution of error detection mechanisms is provided in Figure 8.1, below.

With respect to the interaction between error type and error detection mechanism, a number of operationally relevant findings were obtained. Overall, there was a significant difference in error detection, as a function of error type $\chi2(12, 677) = 173.882$, p < .001. Analysis of the adjusted standardized residuals indicated significant interaction between specific types of errors, and patterns of error detection. The differences in error detection mechanism as a function of error type is presented in Figure 8.2, below.

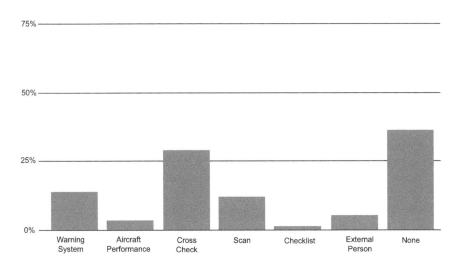

Figure 8.1   Error detection – all errors

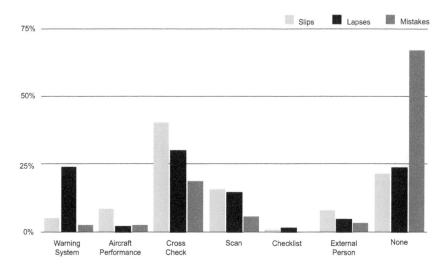

Figure 8.2   Error detection – as a function of error type

Slips, active errors committed in the execution of tasks, were more frequently detected through noticing changes in aircraft status or performance, or through cross-checking. Lapses, failure to execute a planned action due to failures of memory, were more frequently detected by Aircraft Warning Systems. Mistakes, errors that occur during the planning of actions, remained undetected significantly more frequently than other types of error.

Discussion

The results of this study provide new insights into the landscape of error and error detection during normal flight operations. From a general perspective, this study provides some operational examples of resilience engineering in practice. The research findings provide a practical example of how variability in performance is tolerated within a complex socio-technical system such as commercial aviation.

This study demonstrates a system that anticipates error, and has already built into the system multiple components, working both in parallel and in series, in order to assist in the detection of inevitable error. This study has also demonstrated that error tolerance is not as simple as a defences-in-depth approach. The traditional defences-in-depth approach constructs multiple layers of defences that operate in series. If one layer of defence fails, additional layers designed into the system eventually stop the trajectory of error. However, this study has demonstrated how defences must also act in parallel – in order to be adaptive to the stochastic process of error generation.

Stochastic Processes Require Organic Defences

The results of this research also highlight the inherent variability in the processes of error occurrence and error detection. These findings are in line with the current theoretical perspectives of resilience engineering that specifically constructs human error as natural variability in human performance (Dekker, 2003; Hollnagel, 2002; Hollnagel et al., 2006).

Error, when seen in terms of human performance variability within the complex socio-technical system, is stochastic in nature. It is frequently unpredictable, and as such demands a composite approach to risk mitigation. To this end, highly deterministic approaches to error tolerance, which build defences in depth as protective layers built in serial, may not be the most effective approach. Resilience engineering, in contrast, must continue to adopt organic approaches that build defences in parallel, in order to defend against the naturally complex trajectories of error.

Error as Adaptive Mutation

One slightly more complex construction of error as performance variability could suggest that error is a form of behavioral mutation. As within all complex biological systems, mutation is a natural source of evolutionary development and therefore critical for adaptation. Recent theoretical and experimental genetics suggest that adaptation involves progressive shifts towards a phenotype that best fits the present environment of the organism. The conventional thought with respect to this adaptation suggests that such enhancement of an organism's "fitness" is derived from largely random mutations at the genetic level (Orr, 2005).

Mutation is often seen as a negative construct, in a similar way to our traditional views of human error. However, relatively few mutations are detrimental to an organism. The vast majority of mutations are neutral in terms of consequence upon an organism's fitness. More importantly, a proportion of genetic mutations are indeed of significant benefit. Recent genetic science suggests that in mutagenesis experiments, up to 15 percent of mutations are advantageous (Eyer-Walker and Keightley, 2007).

If we consider human error to be a process of variation akin to the genetic variability inherent in mutation, it is possible to see that some error may well be advantageous to performance and therefore adaptive in nature. We already know that some error is deleterious to performance and safety, whereas a significant proportion of error is neutral. The next logical step is to accept that, indeed, some error must have adaptive benefits. To this end, in systems that attempt to standardize and constrain performance to eliminate error, are we simply interfering with natural processes designed to optimize performance?

Modern medicine, which has long accepted genetic variability as largely neutral or advantageous in nature, adopts an approach that tolerates mutation as natural and puts in place processes to identify mutations that are potentially detrimental and then mitigate their consequences on the quality of life of the individual. This approach is remarkably close to the philosophy of threat and error management, and accordingly, we can learn much from the construction of error as potentially adaptive behavioral mutation.

In conclusion, this study has highlighted that resilience engineering operates in practice already within complex systems such as commercial aviation. The next steps in building upon this foundation appear to reside in continuing to examine the stochastic nature of human performance variability, and the design of parallel defences that enable more sophisticated anticipation and detection of error.

References

Dekker, S.W.A. (2003). Illusions of explanation: a critical essay on error classification. *International Journal of Aviation Psychology*, 13(2), 95–106.

Eyer-Walker, A. and Keightley, P.D. (2007). The distribution of fitness effects of new mutations. *Nature Reviews: Genetics*, 8, 610–18.

Helmreich, R.L. (2000). On error management: lessons from aviation. *British Medical Journal*, 320, 781–5.

Helmreich, R.L., Klinect, J.R., and Wilhelm, J.A. (1999). Models of threat, error and CRM in flight operations. In: R.S. Jensen (ed.), *Proceedings of the Tenth International Symposium on Aviation Psychology*, (pp. 677–682). Columbus, OH: Ohio State University.

Helmreich, R.L., Merritt, A.C., and Wilhelm, J.A. (1999). The evolution of crew resource management. *International Journal of Aviation Psychology*, 9(1), 19–32.

Hollnagel, E. (2002). Understanding accidents – from root causes to performance variability. *Proceedings of the IEEE 7th Human Factors Meeting*, (pp. 1.1–1.6). Scottsdale Arizona, 2002.

Hollnagel, E., Woods, D.D., and Levenson, N. (2006). *Resilience Engineering: Concepts and Precepts*. Aldershot: Ashgate.

Klinect, J.R. (2002). LOSA searches for operational weaknesses while highlighting systemic strengths. *International Civil Aviation Organisation (ICAO) Journal*, 57(4), 8–9, 25.

Klinect, J.R., Wilhelm, J.A., and Helmreich, R.L. (1999). Threat and error management: data from line operations safety audits. In: R.S. Jensen (ed.), *Proceedings of the Tenth International Symposium on Aviation Psychology*, (pp. 683–88). Columbus, OH: Ohio State University.

Orr, H.A. (2005). The genetic theory of adaptation: a brief history. *Nature Reviews: Genetics*, 6, 119–27.

Reason, J. (1987). A framework for classifying errors. In: J. Rasmussen, K. Duncan and J. Leplat (eds), *New Technology and Human Error*, (pp. 5–14). Chichester: John Wiley and Sons.

Reason, J. (1990). *Human Error*. Cambridge: Cambridge University Press.

Reason, J. (1997). *Managing the Risks of Organizational Accidents*. Aldershot: Ashgate.

Shappell, S.A. and Wiegmann, D.A. (2000). *The Human Factors Analysis and Classification System – HFACS*. Washington, DC: Federal Aviation Administration – Office of Aviation Medicine.

Shappell, S.A. and Wiegmann, D.A. (2001). Applying reason: the human factors analysis and classification system (HFACS). *Human Factors and Aerospace Safety*, 1(1), 59–86.

Thomas, M.J.W. (2004). Predictors of threat and error management: identification of core non-technical skills and implications for training systems design. *International Journal of Aviation Psychology*, 14(2), 207–31.

Wiegmann, D.A. and Shappell, S.A. (2001). Applying the human factors analysis and classification system (HFACS) to the analysis of commercial aviation accident data. *Proceedings of the 11th International Symposium on Aviation Psychology*. Colombus, OH: The Ohio State University.

Wreathall, J. and Reason, J. (1992). Human errors and disasters. *Proceedings of the 1992 IEEE Fifth Conference on Human Factors and Power Plants*, Monterey, California, (pp. 448–52). New York: IEEE.

Chapter 9

The Role of GPS in Aviation Incidents and Accidents

Gemma Stański-Pacis

Maastricht University, The Netherlands

Alex de Voogt

American Museum of Natural History, USA

Introduction

Interactions between humans and computer based technologies have increased to a great degree as a result of a constant demand for improved efficiency and safety. Simplifying tasks and cost reductions have been the main incentives for implementing automated systems (Parasuraman and Riley, 1997, p. 232; Parasuraman, 2000, p. 246). Research has revealed that the use of automated systems has changed the basic characteristics of the cognitive demands and responsibility of the human operator of systems (Parasuraman and Riley, 1997, p. 232; Bainbridge, 1983, pp. 775–6).

The introduction of automated systems has resulted in unforeseen consequences when human interaction is involved (Parasuraman, and Riley, 1997, p. 231). For instance, the goal of the flight management system is to aid the pilot in navigation, flight planning and aircraft control functions. Practice has shown that this system tends to make low-workload phases of flight easier, such as straight and level flight or a routing climb. On the other hand, the flight management system tends to put more strain on the pilot during high-workload phases, such as maneuvers in preparation for landing (Wickens, Lee, Liu and Becker, 2004, p. 427). Designers develop a system with a particular goal in mind, but may overlook the potential effects of the user's interaction with the system (Bainbridge, 1983, p. 777). Therefore, it is important to consider three parts of human/computer interaction when designing a system.

The first part relates to the purpose of the system. The purpose is expressed in the language of the application domain. For example, a navigational system for aviation makes calculations and aids in navigating to a location. The features of the system are the second aspect. They make it possible for the system to achieve its goal by representing the information visually to the user, for example in the case of navigation. The last component is the user who uses a concept from a subjective point of view when using a system to achieve a goal (Casaday, 1996, pp. 363–4). To predict the effects of an action, the user produces a mental model (Norman, 1998, p. 38). This clearly shows that an interface between machine and user must be designed in a predictable way so that the user forms a clear mental model of the system and of the outcome represented by the system. To create a transparent process, the user must stay involved in the process, and understand the capabilities and limitations of the system, which will also prevent the user from relying too much on the system (Billings, 1997, p. 241). Inagaki (2003, p. 10) supports this statement by looking at several examples where safety may be degraded when humans do not comprehend the intent of automation, or when goals of human and machine are in conflict. In his article, Inagaki (2003) takes a closer look at adaptive automation, which aims at flexibility and at a better synergy between man and machine to improve safety.

In this study the focus is on the use of a GPS system in Aviation in order to reveal underlying factors that result in detrimental flight performance. The approach consists of an analysis of incidents and accidents where GPS was utilized during flight. Investigating human performance in aviation is mainly based on the analysis of incident and accident data. Currently most database reporting systems are not developed around a theoretical framework of human performance. The database systems are employed with the purpose of identifying engineering and mechanical failures. Human performance errors are addressed minimally or are not addressed at all (Wiegmann and Shappell, 2001, p. 1). Although the databases do not include a complete set of information for a human factor specialist, it is still possible to draw conclusions from the analyzed data and indicate the areas for further investigation.

In the long term, it is recommended to integrate the results of incident and accident investigation with other research approaches, for example, cognitive theory, laboratory experiments, simulator studies, field studies, and other empirical studies. Each of these approaches can complement the other and extract specific recommendations for the appropriate design of automation or progressively change the design into an appropriate design for better human performance (Parasuraman, 2000, p. 932).

The Development of GPS Navigation in Aviation

Navigational aids have progressed from primitive aids such as navigation by the angular measurements of the natural stars, to an advanced technology named GPS (Global Positioning System). GPS was based on radio ranging to a constellation of artificial satellites called NAVSTARs. Angular measurements to natural stars were now replaced with ranging measurements to the artificial NAVSTARs (Parkinson, 1996, p. 3). Initially, GPS was developed by and for the military in the 1970s (Parkinson, 1996, p. 4).

In the 1980s, GPS was introduced to the international civilian community (Federal Aviation Administration, 2007), but civilian users of GPS had to cope with reduced accuracy and statistical accuracy variations induced by SA (Selective

Availability). SA is a setting in which GPS signals are intentionally degraded to prevent the use by military adversaries (Federal Aviation Administration, 2007).

In 2000, the U.S. department of Defense turned off SA, giving all GPS receivers improved accuracy (Klepczynski, GPS TAC/WAAS Team, Powers, Douglas and Fenton, 2001).

In 1990, the first GPS receivers especially designed for the General Aviation market became available. They contained databases with geographical information and performed typical area navigation functions as great circle range and bearing, ground speed, and more. Compared to earlier navigation equipment, the pilot benefited from improved accuracy, quicker responses, more accurate velocity measurements and no geographic gaps in the coverage (Eschenbach, 1996, p. 384).

The primary function of a GPS receiver in a cockpit is to enhance the pilot's awareness and to provide information for navigational guidance. During different phases of flight, which are usually separated into "en route", "terminal", "approach" and "landing", the pilot is confronted with different requirements for each of those phases of flight (Eschenbach, 1996, p. 385). Eschenbach (1996, pp. 386–7) mentioned three critical factors that need to be considered when utilizing GPS equipment. First, GPS availability, in which outage of the satellite or equipment failure needs to be taken into account; second, the reliability of GPS with regards to the continuity of the service; and third, GPS integrity in terms of undetected faults. GPS alone cannot deliver the necessary integrity requirements and needs a supplementary system, such as RAIM (Receiver Autonomous Integrity Monitoring), which can act as an integrity monitoring system (Polkinghorne, 1996, p. 21). Furthermore, the pilot has to handle other cockpit demands and GPS receivers must facilitate the flow of information from the pilot to the navigation system and return the flow of information from the navigation system to the pilot (Eschenbach, 1996, p. 387).

Due to the increased demand for GPS in aviation, several manufacturers have put non-standardized GPS units on the market (Wreggit and Marsh II, 1996, p. 1). As seen in Figure 9.1 to Figure 9.6, several types of handheld GPS devices are marketed. They are available in several shapes, with several methods for data entry and retrieval and with several display types.

GPS handheld devices are commonly used by General Aviation pilots, because of their placement possibilities in the cockpit and their affordable prices (Wreggit and Marsh II, 1997, p. 1). The available GPS units have a tendency to neglect human factors principles, which results in non-user-friendly devices (Adam, Deaton, Hansrote, and Shaikh, 2004, p. 114; Wreggit and Marsh II, 1997, p. 14). Research of Heron, Krolak and Coyle (1997, p. 16) and Heron and Nendick (1999, p. 220) support this statement and their study points to flaws in the design of the GPS devices. They have focused on GPS IFR and concluded that the currently marketed GPS receivers trigger human factor problems that manifest themselves in incidents and, without intervention, may result in accidents. These problems affect sensory, perceptual, cognitive and response preparation resources that increase processing workload. This will be reflected in increased workload when executing

Figure 9.1   Garmin 696

Figure 9.2   Garmin 496

Figure 9.3   AvMap EKP-IV

Figure 9.4   AvMap GeoPilot II

Figure 9.5   Garmin 96C

Figure 9.6   Lowrance AirMap 600C

flight tasks (Heron, Krolak and Coyle, 1997, p. 10). For instance, information processing during navigation of a non-precision approach is high. The pilot has to perform his or her tasks in a complex multi-tasking situation with interfering factors, for example when ATC requests the pilot to land on an allocated runway (Heron, Krolak and Coyle, 1997, p. 16). When the pilot is under stress or fatigued and/or dealing with an emergency situation, a non-optimal design of the interface increases the workload. All these factors result in problems concerning information processing and lead to head-down time and loss of situation awareness (Heron, Krolak and Coyle, 1997, p. 16; Heron and Nendick, 1999, p. 202). This raises the question whether the designers of the GPS device have taken into account the variables of human perception and information processing and translated this into the final design of their GPS product.

In General Aviation there are owners and renters. The owners have the advantage of regularly using the same GPS equipment, contrary to renters (Adams, Adams, and Hwoschinsky, 2001, p. 5). This creates problems when a (General Aviation) pilot is accustomed to the use of a particular GPS unit and has to use another GPS unit in another flight (Adam, Deaton, Hansrote, and Shaikh, 2004, p. 110). The difficulty is with the non-transparent GPS interface. It performs complicated calculations, but the user only sees the end result (Graham, 1999, p. 100), which can give the impression that the GPS device is a "walk-up-and-use" device. Casner (2004, p. 10) demonstrated that average pilots flying IFR and using GPS were not proficient for six instrumental flying skills after five ground and five flight learning sessions. This supports the statement that a GPS device should not be perceived as a "walk-up-and-use" system.

Pilot/computer interface issues with GPS units are not limited to this kind of technology. It is just another example of implementing new technology in the cockpit, along with the problems of bringing a product rapidly to the market. Requirements of design are based on engineering and marketing interests rather than human performance interests (Wreggit and Marsh II, 1997, p. 1). With the introduction of new technology, in this case GPS, there are new cognitive demands and performance bottlenecks surfacing (Graham, 1999, p. 99). A study of Wiggins (2007, pp. 259–61) supports this statement. He presented an analysis of pilots who were using GPS as primary navigation tool and their decision-making strategy during flight in certain weather conditions. It became apparent that pilots alter their decision-making strategy and take more risks when flying with GPS as main navigational aid in detoriating weather conditions. Heron, Krolak and Coyle (1997, p. 1) also emphasize that with the introduction of GPS technology the pilot has to cope with new technology and this will show a shift in aviation tasks and will, as Heron, Krolak and Coyle (1997, p. 1) have pointed out, "change the art and craft of flying".

GPS applications in aviation are still evolving. Its utilization for collision avoidance and landing systems is anticipated and will enhance autonomous flight (Eschenbach, 1996, p. 393). Although navigating with GPS is an advanced technology, there are no mandatory training requirements for GPS knowledge, and thus the General Aviation pilot is not obliged to follow a formal training program (Polkinghorne, 1996, p. 21). Results of the study of Dornan, Beckman, Gossett and Craig (2007, p. 6) indicate that a formal training intervention is an effective way to improve the capability of the pilot in using the GPS system in a competent way. Adams, Adams and Hwoschinsky (2001, p. 2) have done a preliminary study of incident and accident analysis, where data was used from the ASRS (They have analyzed 58 incidents and accidents, from 1990–1999, where only 10 reports were related to GPS use.) and the Pilot Deviation database (PD), and they recommend as well that training is a necessity. Adam, Deaton, Hansrote and Shaikh (2004, p. 129) stated that education is the pilot's own responsibility, but that formal training is necessary to achieve it.

The difficulty of extracting GPS-related human machine interaction problems has prevented the identification of the relevant issues in today's aviation. The combined incident and accident study presented here should distinguish the issues that initially need our attention and may prevent an increase in GPS-related incidents and possible accidents. On the basis of the theory on human/machine interaction and the previous studies on GPS utilization in aviation, it is expected that incidents and accidents reveal issues of training and system design. Incident analysis should aid in the prevention of accidents, particularly since incidents can be considered a precursor of an accident and may develop into an accident without intervention.

Method

Data

The data include incident and accident reports of pilots in Commercial and General Aviation and are limited to those who used GPS equipment.

Incident reports were taken from the Aviation Safety Report System (ASRS) database report set. The Global Positioning System (GPS) report set features 50 incidents involving the use of GPS devices for the period 2004 through 2006. This report set can be found on the ASRS website: <http://asrs.arc.nasa.gov/search/reportsets.html>. The ASRS collects, analyzes and responds to aviation safety incident reports that have been submitted voluntarily by pilots, controllers and others. Due to the voluntary submission of these reports it has a reporting bias and represents only a certain group of pilots.

Accident reports for the period 1993 through 2007 from the National Transportation Safety Board (NTSB) database were collected from their online database. The data were extracted from: <www.NTSB.gov/NTSB/Query.asp>. The NTSB database contains information about all aviation accidents for which a summary format is issued. All accidents are reported and there is no significant reporting bias compared to the voluntary submitted ASRS reports.

Procedure

The incident reporting set is not available in a suitable format for analysis. The narratives were read in detail in order to make an decide whether GPS utilization was a factor in the incident. The extracted events, where GPS played a role in the incident, were classified and particulars of the incident were scored. After scoring and classifying the events, the events were clustered and coded for later analysis.

Not all incident reports contained information regarding the GPS device that was used. The age of the pilot and the number of flight hours were neither available in the incident report set. Furthermore, the extracted data from the narrative is vulnerable to subjective interpretations and may be scored differently by another researcher. For example, in ASRS report 703731 the pilot explained that the high workload increased when using GPS compared to when using no GPS system. In this case the report was assessed and classified under "not being well trained or knowledgeable with GPS system". Another interpretation could be that the pilot did not yet have extensive flight hours and that he found the workload too high, because the pilot had to execute simultaneous other flight tasks. In the absence of data on flight experience, the first interpretation is chosen.

A follow up study analyzed accident reports. A total of 2085 accident reports were searched and probable cause statements were collected. As in the case for the incident report set, the formats of the accident reports were not immediately suitable for analysis. All 2085 reports had to be read in detail in order to assess if the report was going to be used for analysis. The reports that were included in the analysis contain the term "GPS" or "Global Positioning System" in the probable cause file, the reports containing only "navigation" or "nav" were left out. If the report contained the term "GPS" or "Global Positioning System" the particulars of the accident were coded, clustered and used for analysis. After assessing all 2085 accident reports the number was narrowed down to an accident report set of only 29 cases.

The data of the accidents reports do not have the exact same logging format as the incident reports and therefore the data are scored and classified differently. As in the case of the incident reports, the analysis of the accident reports is also vulnerable to subjective interpretations. For example, in the NTSB report NYC00FA209 the assessment was made that GPS was used. It was concluded that the pilot was lost or disoriented, but another researcher could interpret this differently and classify it, for instance, under decision making error. The pilot decided to fly in night time with bad weather conditions and from the accident report no assessment was possible about the IMC flight hours and night flight hours of the pilot.

Extracting details from the accident report about the experience of the pilot with the GPS system is not possible. The data from these accident reports was not suitable for more detailed extraction of the pilot-system interaction.

Data Analysis

The data of the incidents were analyzed and scored based on the following points: date/day/local time; flight conditions; light conditions; weather; flight phase/flight phase status; function crew; make/model airplane; altitude; state; local reference; regulation flight conducted under; factors/problems; events; experience of the pilot using the GPS device; presence of a GPS device.

The distribution of the occurrences of the incidents during the period 2004–2006 per month and weekday were analyzed. The flights during day and night were compared and the frequency of the flight condition was counted. An analysis of the problem area established if flight crew performance, that is, human error, played a role in the incident.

The data of the accidents are analyzed and scored based on the following points: date/local time; make model; engine model/number of engines; aircraft damage; type of flight operations; regulation flights conducted under; severity of accident; flight conditions; light conditions; ceiling; visibility; obscuration; pilot age; pilot certificate/rating; instrument rating; pilot flight hours; occurrence; phase of operations; factor findings/cause.

The frequency of the accidents for the period 1993–2007 was analyzed to determine a trend in the occurrence of the accidents. The clustered factors were analyzed to establish what factor occurred most frequently.

Results

Incidents

From the 50 incident reports, five incidents were excluded from the data. GPS utilization could not be analyzed as a factor in the incident and therefore these five incident reports were excluded from the analysis.

The flight crew performance was, in 40 cases, assessed as a contributing factor to the incident. For example, in the ARSR report 694032 a pilot flies into military class D area because he did not monitor his position.

In four cases the air traffic control performance was assessed as a contributing factor to the incident. The contributing role of air traffic control was mentioned in only one incident. ASRS incident report 672927 shows that the tower controller forgot to set his equipment correctly, a manual switch that determined the approach direction. As a result the pilot received wrong information about the approach.

The extracted factor "insufficient GPS knowledge/no training" occurred in 21 of the studied incidents and occurred more frequently than any other factor (see Figure 9.7).

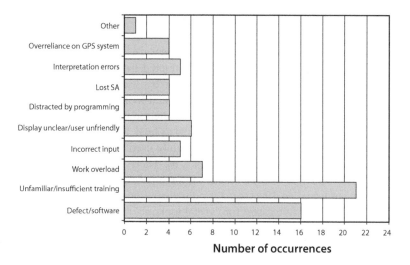

Figure 9.7   Events involving GPS

Incidents in which the pilot reported to be experienced with GPS were compared to those in which the pilot was unfamiliar with the GPS system or only recently had had some training but was not yet accustomed to the navigation system. The level of experience of the pilot was related to their flight being catagorized as General Aviation or Commercial Aviation. The results indicate that pilots who were not in General Aviation were more likely to have had experience with the use of GPS systems. There was a significant difference between an experienced GPS user and a non experienced GPS user since the latter was more likely to be in the field of General Aviation ($\chi^2 = 8,259$; $p < 0,05$; df = 1).

A second comparison was made between the level of experience with GPS and the flight phase in which the pilot was flying. The results show that an experienced GPS user more often encountered a problem during the non-descent phase than during the descent phase, contrary to the non-experienced GPS user when problems more often arose during the descent phase. An experienced GPS user appears to encounter problems in 73.1 percent of the reported cases of non-descent, while in the descent phase the non-experienced GPS user encounters most of the problems (73.7 percent). The difference was significant between the level of experience with GPS and the flight phase ($\chi^2 = 9,644$; $p < 0,05$; df = 1).

Accidents

In the period between 1980 and 1993 no accidents were reported in which a GPS or global positioning system was mentioned in either the factual or probable cause report. The analysis of the accident reports between 1993 and 2007 showed that, in 29 cases, it could be inferred that the accident was related to the utilization of the GPS device. In 2001 five accidents occurred, while in the years 1993, 1995, 1997, and 2006 only one such accident occurred. There was no clear trend in the occurrence of accidents where GPS equipment was used.

The following clusters of factors were extracted from the data: "failure with GPS connection"; "GPS not adequately used by user"; "disorientation/lost"; "diverted attention"; "hardware issues"; "no visual lookout". The highest rate of accidents, seven, involved the factor "diverted attention" due to GPS use.

In accidents where the injury was classified as fatal, the number of injuries was highest in the events of "GPS was not adequately used" (N = 6); "disorientation/lost" (N = 6); and "no visual lookout" (N = 6). There was one fatal accident in which diverted attention was an issue. Minor injuries were reported for "GPS hardware or software" in 10 cases; for "disorientation or lost" in six cases; for "diverted attention" in five cases; for "visual lookout" there was only one case.

In two reports (NTSB report LAX01FA055 and SEA01LA138) it was mentioned in the probable cause statement that GPS was a factor in the accident. In all other cases, the narrative text had to be used to identify the use and the problem with the use of the GPS.

Overviews are given in Figure 9.8 of GPS-related factors in relation to the average number of flight hours of the pilot.

There were 46 accidents where GPS was used and weather conditions were a factor. These reports were excluded because no clear inference could be made whether GPS use was a contributing factor in the problem.

No additional significant relationships in the data were found, which is partly due to the low number of cases in the dataset.

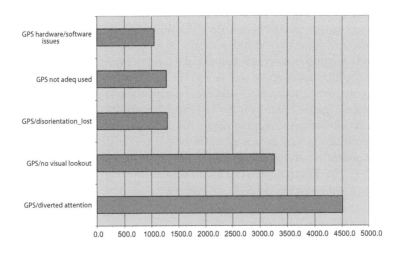

Figure 9.8   Average number of flight hours and GPS-related factors

Discussion

The purpose of automation is to relieve the workload or simplify the flight tasks of the pilot during flight. Introducing new technology is not without risks. Research has shown that utilization of new technology can result in undesired results and is affected by the way the pilot uses new technology. The GPS device is a relatively new technology and has been introduced in General Aviation without formal regulations for the design of the GPS system and without any training requirements for the use of the GPS system.

The results from the incident analysis study indicate that the expected problems from the theoretical literature and from previous research on human/machine interaction are materializing in the GPS environment.

In 40 incident cases human performance was at the heart of the problem. The level of experience with the GPS system plays a main role. The data support the view that pilots in General Aviation experience more problems with the utilization of the GPS system, since the pilots are not obliged to undergo formal training. The level of experience played a role in the flight phase. Novice GPS users experience a problem when utilizing a GPS system during the descent phase, since this carries a high workload. The unexpected part of the data analysis is the fact that a more experienced GPS user has an issue when utilizing a GPS system during the non-descent phase. Future study may explain this particularity.

The accident data suggest that incidents are indicators of problems that can lead to an accident. The data show that compared to the incident data there are similar issues that surface in the accident data. The analysis of the accident data indicates that pilots with a higher average of flight hours have an issue in the area of diverted attention when utilizing GPS and pilots with the lowest average of flight hours have an issue with the "hardware/software".

Furthermore, the average age of the pilot who had an accident when utilizing GPS during flight is approximately 52 years, and seems to indicate that this group needs more assistance or knowledge in how to utilize their GPS system. The statistics of the 2003 NTSB (2006) annual review of General Aviation aircraft accidents report that the average age of General Aviation pilot in 1994 was 41 years and by 2003 it was 45 years old. Statistics provided in the 2007 report from GAMA (General Aviation Manufacturers Association, 2007) show the average age of a General Aviation pilot has remained stable from 2004 through 2007 with an average age of approximately 45 years. These statistics indicate an aging group with a slower learning curve and a group that may not be that well accustomed to the fast evolving technology as the new generation of pilots.

This study has been limited by the details provided in the incident and accident reports that do not always allow conclusions concerning the role of GPS during the accident. As mentioned earlier the databases are not specifically designed for human performance analysis and may render ambiguous results. The reports are sensitive to subjective interpretations and inconsistencies. The advantage of the ASRS incident reports the additional details and clarification given by the pilot when interviewed about the incident. The accident reports do not have this information making it difficult to assess how the pilot perceived the utilization of the GPS system during flight. It is recommended that the experience or knowledge level of the pilot with GPS is included in both the database of the incidents and that of accidents.

In the NTSB accident report no clear inference were possible about the role of GPS use during deteriorating weather conditions and these reports were not further analyzed in this study. Indirect inferences can be made that using GPS has a psychological influence on what decision-making strategies are applied.

Another limitation is found in the missing details from the incident and accident report about the user-friendliness of the GPS equipment. The inclusion of this kind of information may provide better recommendations regarding the design of the GPS equipment and the interaction that occurs with the pilot.

Although GPS navigation is a useful tool in aviation, it also introduces problems in human/machine interaction. It is well known in the literature that this interaction needs to be addressed in order to minimize both incidents and accidents in the future. With the perspective that GPS technology is still evolving there needs to be a focus now on regulating the introduction and implementation of new designs, particularly for the market of General Aviation. Furthermore, training has to be regulated before utilizing GPS equipment in General Aviation flights.

The results of this research confirm that focusing on standardization of GPS devices (Joseph and Jahns, 2000, p. 6) and enhancing GPS training should mitigate the problems related to GPS use in aviation. Educating the pilot will make the pilots aware of their own responsibility for safe GPS utilization and teach them what the capabilities and limitations are of a GPS system during flight.

References

Adam, H., Deaton, J.E., Hansrote, R.W. and Shaikh, M.A. (2004). An analysis of GPS usability and human performance in aviation. *International Journal of Applied Aviation Studies*, 4, 107–31.

Adams, C.A., Adams, R.J. and Hwoschinsky, P.V. (2001). Analysis of adverse events in identifying GPS human factors issues. Volpe NTSC. (See also the 11th International Symposium on Aviation Psychology, March 5–8, 2001, Columbus, Ohio. <http://techreports.larc.nasa.gov/ltrs/PDF/2001/mtg/NASA-2001-11isap-caa.pdf>).

Aviation Safety Reporting System. (2007). ASRS database report set global positioning system (GPS) reports: <http://asrs.arc.nasa.gov/search/reportsets.html>.

Bainbridge, L. (1983). Ironies of automation. *Automatica*, 19, 775–9.

Billings, C.E. (1997). *Aviation Automation: The Search for a Human-Centered Approach*. Mahwah, NJ: Lawrence Erlbaum Associates Inc.

Casaday, G. (1996). Rationale in practice: templates for capturing and applying design experience. In: T.P. Moran and J.M. Carroll (eds), *Design Rationale: Concepts, Techniques, and Use*. New Jersey: Lawrence Erlbaum Associates Inc.

Casner, S. (2004). Flying IFR with GPS: how much practice is needed? *International Journal of Applied Aviation Studies*, 4, 81–97.

Dornan, W.A., Beckman, W., Gossett, S. and Craig, P.A. (2007). A FITS scenario based training program enhances GPS pilot proficiency in the general aviation pilot. Retrieved November 23, 2008, from <http://www.faa.gov/education_research/training/fits/research/media/FITS_for_GPS-2007.pdf>.

Eschenbach, R. (1996). GPS application in general aviation. In: B.W. Parkinson J.J. Spilker Jr., P. Axelrad and P. Enge (eds), *Global Positioning System: Theory and Applications Volume II*, (pp. 375–95). Washington, DC: The American Institute of Aeronautics and Astronautics, Inc.

Federal Aviation Administration. (2007). Evolution of the United States national airspace: the move towards performance based navigation. Retrieved April 21, 2008, from FAA: <http://www.faa.gov/about/office_org/headquarters_office/ato/service_units/techos/navserivces/history/satnav/index/cfm>.

General Aviation Manufacturers Association. (2007). General aviation statistical databook and industry outlook. Retrieved December 7, 2008, from GAMA: <www.gama.aero/events/air/dloads/2007GAMADatabookOutlook.pdf>.

Graham, J.M. (1999). The impact of new technology on general aviation: global positioning system receivers, the Federal Aviation Administration, and the small aircraft pilot. *Proceedings of the 1999 IEEE International Symposium on Technology and Society on Women in Technology: Historical, Societal and Professional Perspectives*, (pp. 99–104).

Heron, R.M., Krolak, W. and Coyle, S. (1997). *A* human factors approach to the use of GPS receivers. *Proceedings of the 9th Canadian Aviation Safety Seminar*. Calgary, Alberta, Canada. (See also Bluecoat Digest, <http://bluecoat.org/reports/>).

Heron, R.M. and Nendick, M.D. (1999). Lost in space: warning, warning, satellite navigation. In: D. O'Hare (ed.), *Human Performance in General Aviation*, (pp. 193–224). Aldershot: Ashgate.

Inagaki, T. (2003). Adaptive automation for comfort and safety. *International Journal of ITS Research*, 1, 3–12.

Joseph, M.K. and Jahns, D.W. (2000). Enhancing GPS receiver certification by examining relevant pilot-performance databases (Rep. No. DOT/FAA/AM-00/4). Washington, DC: Office of Aviation Medicine.

Klepczynski, W.J., GPS TAC/WAAS Team (and-730), Powers, E., Douglas, R. and Fenton, P. (2001), ION (the Institute of Navigation) newsletter 11, 2.

National Transportation Safety Board. Annual review of aircraft accident data U.S. general aviation, Calendar year 2003. (Publication No. ARG-07-01, adopted on November 29, 2006). Retrieved December 07, 2008 from NTSB: <http://www.ntsb.gov/publictn/2007/ARG0701.pdf>.

National Transportation Safety Board online accident database. Retrieved July, 2007 from: <http://www.ntsb.gov/ntsb/query.asp>.

Norman, D.A. (1998). *The Design of Everyday Things*. New York: Basic Books.

Parasuraman, R. (2000). Designing automation for human use: empirical studies and quantitative models. *Ergonomics*, 43, 931–51.

Parasuraman, R. and Riley, V. (1997). Humans and automation: use, misuse, disuse, abuse. *Human Factors*, 39, 230–53.

Parkinson, B.W. (1996). Introduction and heritage of NAVSTAR. In: B.W. Parkinson J.J. Spilker Jr., P. Axelrad and P. Enge (eds), *Global Positioning System: Theory and Applications Volume II*, (pp. 375–395). Washington, DC: The American Institute of Aeronautics and Astronautics, Inc.

Polkinghorne, A. (1996). *The Trouble with GPS*. Civil Aviation Safety Authority, Australia, 1, 20–21.

Wickens, C.D. Lee, J.D., Liu, Y. and Becker, S.E.G. (2004). *An Introduction to Human Factors Engineering*. NJ: Pearson Education Inc.

Wiegmann, D.A. and Shappell, S.A. (2001). A human error analysis of commercial aviation accidents using the human factors analysis and classification system (HFACS) (Rep. No. DOT/FAA/AM-01/3). Washington, DC: Office of Aviation Medicine.

Wiggins, M. (2007). Perceived GPS use and weather related decision strategies among pilots. *International Journal of Aviation Psychology*, 17, 249–64.

Wreggit, S.S. and Marsh II, D.K. (1996). Cockpit integration of GPS: initial assessment-menu formats and procedures (Rep. No. DOT/FAA/AM-98/9). Washington, DC: Office of Aviation Medicine.

Chapter 10

Creating Safer Systems: PIRATe (The Proactive Integrated Risk Assessment Technique)

Brenton Hayward, Andrew Lowe and Kate Branford
Melbourne, Australia

Introduction

Major aircraft accidents involving multiple fatalities are typically followed by intensive investigations to identify and address the range of factors that contributed to the event. Contemporary examples of this are the 2001 Milan Linate runway collision, mid-air collisions over the German town of Überlingen (2002) and the remote rainforests of Brazil (2006), the 2007 Yogyakarta and Congonhas-São Paulo runway overrun accidents, the 2008 crash on take-off at Madrid Barajas, and the loss of Air France Flight 447 over the Atlantic Ocean in 2009. Such events are tragic for those directly affected and provoke strong emotional reactions within the aviation community and broader society. This has included a disturbing trend in recent years towards increasing criminalization of the front-line aviation workers associated with the events.

Safety investigation methods for learning from and preventing recurrence of these accidents have evolved considerably over the past 20 years. One key improvement has been the transition to systemic safety occurrence analysis methods. Examples of these include Tripod (Doran and van der Graaf, 1996; Hudson et al. 1994), the Incident Cause Analysis Method (ICAM; BHP Corporate Safety, 2000), AcciMap (Branford, Naikar and Hopkins, 2009; Hopkins, 2000; Rasmussen, 1997), the Human Factors Analysis and Classification System (HFACS; Wiegmann and Shappell, 2003) and the Systemic Occurrence Analysis Methodology (SOAM; EUROCONTROL, 2005). While these techniques have their differences, they share a number of common attributes:

- They are evolutions of, or at least consistent with, the work of Reason (1990, 1997, 2008) in that they focus on identification and control of the broader systemic factors that contributed to an accident or incident, rather than just the behavior of operators. They begin by identifying the immediate precursors to the accident and extend to consider the contextual, environmental and organizational factors that promoted, allowed, or failed to prevent the negative outcome. In considering these "latent conditions" (Reason, 1991), the activities of frontline operators are linked with the context in which they took place, to enable a more complete understanding of how and why the event occurred.

- By moving beyond specific, low-level details of accidents and incidents to consider broader organizational and system factors, these techniques enable commonalities between different accidents to be identified and problem areas to be prioritized and addressed.
- The corrective actions that emerge from such techniques focus on the "system", rather than the "person". Recommendations are designed to address the barriers or controls that failed as well as the areas of organizational vulnerability that contributed to the unsafe situation, rather than the behavior of the particular personnel involved in the event. Systemic improvements have the capacity to generate significant reduction in risk throughout the system by addressing hazards and error-producing conditions at their source.
- Given the right circumstances, these systemic methods can facilitate a positive shift from a "blame culture" through "no blame" and towards a "just culture" approach to safety investigation, and to the reporting of "normal errors".

The investigation process concludes with the release of a report that identifies a set of contributing factors and recommends corrective actions to address each of these. The recommendations made are seldom new or unexpected, and may even evoke a sense of *déjà vu* amongst those who take an interest in accident investigation and prevention. The reason for this is that the *contributing factors* in many accidents are already known and understood. People working within the industry are seldom surprised at the findings of investigation, because the same factors have contributed to similar scenarios in the past, either as accidents or as "near misses". What is surprising and frequently disappointing is the fact that the industry has not learned from the prior events and acted effectively to mitigate the risks and prevent similar accidents from happening again.

This situation is not unique to the aviation industry. Operational personnel in most industries possess a wealth of knowledge about the latent conditions that contribute to such tragedies. This chapter discusses a technique that aims to harvest this knowledge and use it to identify and address these conditions *before* their potential for harm and damage is realized.

Safety Occurrence Investigation

Regrettably, the investigation of safety occurrences will remain an inevitable activity in managing risk in aviation and other hazardous industries. Diligent systemic occurrence investigation is essential, as most accidents and incidents provide the opportunity to understand the complete set of conditions that contributed to the outcome and to identify appropriate corrective actions. As a tool for organizational learning and safety risk management, however, investigation processes have obvious shortcomings:

- By definition, investigations are reactive, only being undertaken *after* a safety occurrence or near miss. The more serious of these events often entail terrible costs, in human and other terms, as well as consequential losses such as the damage to an organization's reputation and potentially a threat to its very existence. There is inevitably a high price to pay for the organizational learning derived.
- Accidents and serious incidents typically contain a random element – a highly unusual confluence of actions, events and conditions – that contributes to the fact that the event was not predicted, and therefore not prevented. As a result, investigations can tend to focus in minute detail on those rare and unique conditions that contributed to the particular occurrence. This can lead to undue focus on peripheral issues and may divert attention and resources from other areas of risk within the system.
- The cost of some investigations into high profile accidents can be disproportionate to the benefits obtained. Commissions of Inquiry and other high profile investigations deliver authoritative and exhaustive findings borne of epic legal scrutiny, but at a cost much greater than a competent systemic safety investigation would incur. The inquiry itself and the implementation of its recommendations consume substantial resources that could be applied to more appropriate proactive safety management practices and other important aspects of risk reduction within the industry under investigation.
- Lastly, reactive investigations can only make a limited contribution in ultra safe industries, where accidents and serious incidents – and thus the opportunity to learn – are thankfully rare.

In light of these limitations, it is useful to consider whether alternative approaches exist that can produce the benefits of occurrence investigations, but without some of the associated costs. Current proactive risk management techniques include safety audits and inspections, specialist peer reviews (eg., International Atomic Energy Agency, 2006; World Association of Nuclear Operators, 2005), and LOSA-style "fly-on-the-wall" operational assessments for the airline industry (Line Operations Safety Audit; Helmreich, Klinect and Wilhelm, 2001, 2003; International Civil Aviation Organization, 2002) and in Air Traffic Control (NOSS: Normal Operations Safety Survey; see SKYbrary, 2010). These techniques tend to focus on deviations or gaps at the level of the operational interface, that is, *local* conditions (hazards, etc.), violations or errors that need to be eradicated, corrected or in some way managed. This is based on the reasonable assumption that addressing these deficiencies will be beneficial in reducing the likelihood of an accident. Even those safety assessment techniques that do extend to consider higher level systemic factors, such as safety culture or risk management processes, tend not to illuminate the full range of organizational elements that may contribute to accidents.

The essence of the Reason Model is that local factors combine with higher-level latent conditions to produce an "organizational accident". It would be logical then to audit all of these elements in a holistic way, rather than in isolation. The following describes an approach designed to do just that.

Proactive Integrated Risk Assessment Technique

The Proactive Integrated Risk Assessment Technique (PIRATe) is based on a contemporary adaptation of the principles of Reason's "Swiss Cheese" model of safety occurrence analysis, specifically the Systemic Occurrence Analysis Methodology (SOAM) developed for EUROCONTROL (see EUROCONTROL, 2005; Hayward and Lowe, 2004; Licu, Cioran, Hayward and Lowe, 2007). SOAM includes specific taxonomies and terminology to describe the various factors that contribute to any level of safety occurrence. It prescribes standardized processes for gathering evidence, analyzing contributing factors and ensuring that findings are clearly linked to recommendations, providing guidance to safety investigators and analysts in key aspects of an investigation. A fundamental goal of SOAM is to broaden the spotlight of investigation from a narrow focus on the actions or omissions of front-line workers to include organizational and systemic contributions to the safety occurrence.

Building on these principles, PIRATe is a systemic *proactive* approach for identifying operational risk. It enables *potential* or hypothetical safety occurrences to be analyzed, drawing on the collective experience and knowledge of operational personnel about unsafe conditions and behaviors in their workplace. These risk factors are understood implicitly by frontline operators in safety-critical roles, but are seldom made explicit and are often not apparent to those at higher levels of the organization.

The aim of PIRATe is to elicit this understanding – people's insights into the hazards that confront them each day in their operational environment – and to trace these hazards back to their underlying systemic origins. To this extent, every experienced airline pilot, cabin crew, ramp worker, air traffic controller and maintenance engineer is a local operational safety subject matter expert (SME), possessing valuable, first-hand knowledge about current, everyday hazards and risks and the potential for an accident.

As a derivative of the Reason Model, PIRATe integrates the contributing factors to a hypothetical event into a unique and holistic analysis. Corrective actions flow from the analysis in the same way they do following a competent systemic occurrence investigation. Use of PIRATe involves the following steps:

1. Assemble an appropriate group of SMEs. In the authors' experience, it is productive to use this process as a component of other safety-related

training, for example Crew Resource Management (CRM) courses, where safety issues will be top of mind for participants and they will be primed to transfer ideas into a holistic accident scenario. This can also be done via stand-alone workshops, if preferred.

2. Review and explain the systemic occurrence analysis approach. Participants need to have a good working understanding of the principles of the Reason Model, or a similar systemic occurrence analysis method, to be used as a framework for developing, analyzing and reporting their own scenario.

3. Ask the SMEs to work in small groups to "design their next accident" – to identify the conditions within their system that could contribute to a realistic safety occurrence, using their implicit understanding of local operational hazards, error-producing conditions and absent or imperfect barriers. Groups are encouraged to spend some time in discussion to agree on the broad nature of the accident scenario, which is itself enlightening. For aviation, the scenario might involve aircraft operations, maintenance, ramp, or perhaps a production mishap, depending on the organization(s) involved. After agreeing on a scenario, participants are asked to specify the ingredients (conditions) that are likely to contribute to their hypothetical occurrence.

4. Provide empirical data, where available, to guide both the selection of accident scenarios and the determination of contributing factors. For example, an airline's Flight Operations Quality Assurance (FOQA) data may suggest that an accident associated with an unstabilized approach is worth analyzing. Incident database readouts might suggest particular workplace conditions (fatigue, circadian disruption, equipment or workspace design, crew monitoring or communication deficiencies, etc.) are common contributing factors that might be incorporated into a hypothetical scenario.

5. Ask the SMEs to analyze this hypothetical event, including the organizational factors that are responsible for each hazardous condition or inadequate barrier. They then produce recommended actions to correct each organizational factor and to strengthen safety controls. Custom-designed worksheets can be supplied to facilitate the analysis process and to provide a record of findings and recommendations. It is recommended that facilitators experienced in accident investigation and the use of systemic investigation techniques oversee the process.

PIRATe Application: Examples

PIRATe has been employed successfully to identify operational risk scenarios with groups in the aviation, nuclear power, shipping, rail and hazardous chemical storage industries. Some of these applications and their outcomes are described in further detail below. Examples of the output generated from such an analysis are provided:

1. The process has been used by groups of four to six pilots in more than 200 CRM courses conducted by a major international airline. The pattern of organizational factors appearing across a large number of accident scenarios was analyzed (in an internal study) to identify highest priority areas for potential action. The fact that one high-likelihood, high-consequence and realistic accident was a recurring scenario identified by independent groups of line pilots reportedly influenced the airline's decision to cease operating into the location involved.

2. Groups of operational staff at another major airline were asked to use the approach as part of a human factors and aviation safety management training course. An example of one of the analyses developed during the course is provided in Figure 10.1. The hypothetical accident involved an engine failure on take-off, resulting in impact with terrain and multiple fatalities. The human involvement components related to a mis-declaration of cargo weight; the pilot's failure to recognize the threat of insufficient power and respond appropriately to avoid terrain; and the co-pilot's lack of intervention. The contextual conditions influencing these components are shown, including the factors such as the absence of a backup Load Master, aircraft overloading, maintenance issues, undefined interior airfield obstacles, and co-pilot inexperience. The organizational factors that, in turn, contributed to these conditions are then depicted in the left-most column of the chart, including factors relating to staffing, compliance monitoring, commercial pressures, maintenance deficiencies, risk appreciation and training design. The barriers that failed or were missing and therefore did not prevent the occurrence or limit the severity of its consequences are also depicted. Corrective actions were developed to address the systemic deficiencies that contributed to the outcome (that is, the organizational factors and absent or failed barriers).

3. The process has also been used with helicopter maintenance instructors at an organization in Europe to identify and address training inadequacies with the potential to contribute to maintenance-related safety events. This application of PIRATe led to the identification of a critical gap in the training process, which became the focus of the organization's next human factors training course.

4. Groups of senior managers at a number of nuclear power generation and support facilities in Sweden identified potential safety occurrences in their "ultra-safe" industry, drawing on local and relevant international incidents. Some groups chose to apply the process to a potential security event, identifying contributing factors, organizational deficiencies and inadequate barriers, which could then be addressed.

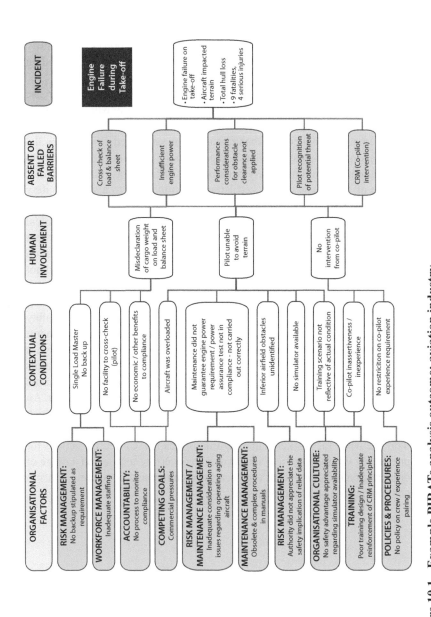

Figure 10.1 Example PIRATe analysis outcome: aviation industry

5. The technique has been used with a major aircraft manufacturer to predict aircraft design and manufacturing problems, and to identify latent conditions with the potential to provoke aircraft maintenance-related occurrences and operator accidents.

6. Ships' crews from a maritime company operating a global fleet of Very Large Crude Carriers (oil "supertankers") employed the technique within Maritime Resource Management (MRM) training courses to identify potential accident scenarios within their sphere of operations, including ship losses, vessel hijackings and environmental disasters.

7. The PIRATe process has been adopted by a European air force for use within aircrew CRM training programs and more recently within a training program for newly appointed squadron supervisory pilots.

8. An international helicopter manufacturer has employed PIRATe for the diagnosis of maintenance related latent conditions with the potential to contribute to operational safety occurrences. PIRATe was introduced after a range of previously unchecked latent conditions led to a serious safety occurrence.

9. An abridged version of the PIRATe process was used at a bulk chemical storage facility to identify organizational and contextual conditions that could potentially result in spillage or contamination events. The application resulted in the identification of threats and errors that management had not previously recognized.

10. In another application, mixed groups of rail safety workers (including train drivers, signallers / controllers, track maintenance workers and safety specialists) analyzed five high-risk accident scenarios involving harm to track workers. These events had been previously identified through a qualitative Risk Assessment. The corrective actions from all five scenarios were collated into a set of recommendations to improve worker safety across the entire track maintenance system for a major suburban rail operator. An example of one analysis from this project is provided in Figure 10.2, which depicts a worker fatality injured as a result of a train passing through a worksite while the worker is in a hazardous location. Corrective actions for preventing this hypothetical scenario from occurring focused on a range of organizational factors, from organizational culture and competing goals to communication and training issues.

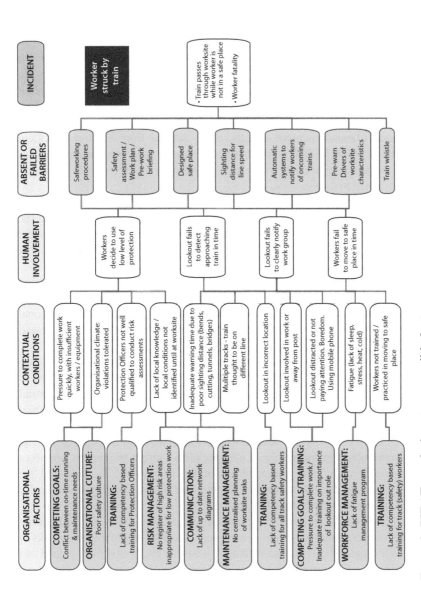

Figure 10.2 Example PIRATe analysis outcome: rail industry

Benefits

There are a number of important benefits associated with PIRATe as a safety enhancement tool:

1. It is a proactive methodology that provides valuable "free lessons" for the organizations and industries concerned. An organization need not experience an accident to benefit from the approach – the outcomes are based on experiences of near misses from operators or from their implicit understanding of areas of the system where the potential for hazardous outcomes exists.

2. It has a sound theoretical basis, being derived from a proven accident analysis tool. Reason's "Swiss Cheese" Model is now a familiar concept in numerous industries, which enables this derivative process to be readily explained to users.

3. It is an integrated technique, combining information from many sources and the ideas from a group of local experts into a holistic, systemic analysis. Most importantly, the technique enables operators to draw on their current experience of, and concern about, actual operational hazards, in a way that other risk reduction techniques do not.

4. Risk Assessment is inherent in the process. As in the examples cited within this paper, highest risk events can be identified, for example from FOQA or other objective data, and combined with the insights of operational personnel with local knowledge and experience. Risk is directly addressed through recommendations made as part of the technique.

5. In addition to the output from PIRATe, there are important secondary benefits from taking part in the process. Participation can be expected to heighten awareness among those involved regarding the nature of industrial accidents – the human involvement (errors and/or violations), contextual conditions and latent organizational contributions, and the barriers that can fail or may be missing – and how people at all levels of an organization can contribute to reducing risk. Like specialized safety training, it promotes wariness about the vulnerabilities of the system and the potential for accidents and incidents, countering the sense of complacency that can arise in industries with good safety records.

6. The approach can facilitate communication about safety issues between those operational personnel with a good understanding of the safety risks in their organizations and management, who may previously have gained this understanding only after a significant safety occurrence. The hypothetical and systems-based nature of the approach promotes confidential, open disclosure among participants regarding the potential for hazardous outcomes, and managers can access these data to reap the benefits of this local knowledge and expertise.

7. There is potential to aggregate the findings from a number of separate PIRATe sessions. For example, the frequency with which similar accident scenarios are appearing can be determined, providing a ballpark measure of "likelihood", and thus risk. A second option is to use a standard framework of Organizational Factors, and to ascertain how often factors such as "Procedures", "Training", "Accountability", or "Safety Culture" arise, and to look at generic remedial action in the highest priority areas. Similarly, each hypothetical accident will identify a number of barriers (defences, controls) that can be compared and strengthened where necessary.

Conclusion

Systemic safety occurrence investigation is essential, and it is important that it is done well, using structured and standardized methods to identify and address the full range of conditions that contribute to an accident or incident. Investigation can, however, be complemented by proactive techniques that provide equal, if not greater, benefit, without significant financial or human cost. PIRATe is a proven proactive approach for identifying and reducing operational risk across a range of industries, now including civil and military aviation, nuclear power, chemical storage, rail and maritime operations. It is considered a useful tool for enhancing organizational learning and helping to ensure that operational safety standards in a variety of settings are continuously improved.

References

BHP Corporate Safety. (2000). Incident Cause Analysis Method Investigation Guide, Issue 1, March 2000. Melbourne.

Branford, K., Naikar, N. and Hopkins, A. (2009). Guidelines for AcciMap analysis. In: A. Hopkins (ed.), *Learning from High Reliability Organisations*. Sydney, NSW: CCH Australia.

Doran, J.A. and van der Graaf, G.C. (1996). Tripod-beta: incident investigation and analysis. In: *Proceedings of the Third International Conference on Health, Safety and Environment. New Orleans*, LA: Society of Petroleum Engineers.

EUROCONTROL. (2005). *EAM2/GUI8: Systemic Occurrence Analysis Methodology (SOAM)*, Edition 1.0. Brussels: Author.

Hayward, B. and Lowe, A. (2004). Safety investigation: systemic occurrence analysis methods. In: K.-M. Goeters (ed.), *Aviation Psychology: Practice and Research*, (pp. 363–80). Aldershot: Ashgate.

Helmreich, R.L., Klinect, J.R. and Wilhelm, J.A. (2001). System safety and threat and error management: The line operations safety audit (LOSA). In: *Proceedings of the Eleventh International Symposium on Aviation Psychology*. Columbus, OH: The Ohio State University.

Helmreich, R.L., Klinect, J.R. and Wilhelm, J.A. (2003). Managing threat and error: data from line operations. In: G. Edkins and P. Pfister (eds), *Innovation and Consolidation in Aviation: Selected Contributions to the Australian Aviation Psychology Symposium 2000*. Aldershot: Ashgate.

Hopkins, A. (2000). *Lessons from Longford: The Esso Gas Plant Explosion. Crows Nest*. NSW: Allen and Unwin.

Hudson, P.T.W., Reason, J.T., Wagenaar, W.A., Bentley, P.D., Primrose, M. and Vissler, J.P. (1994). Tripod delta: proactive approach to enhanced safety. *Journal of Petroleum Technology*, 46(1), 58–62.

International Atomic Energy Agency. (2006). Report: integrated regulatory review service (IRRS), full scope, report to the government of France. Paris: Author.

International Civil Aviation Organization. (2002). Line operations safety audit (LOSA) (Doc 9803 AN/761). Montreal: Author.

Licu, T., Cioran, F., Hayward, B. and Lowe, A. (2007). EUROCONTROL – Systemic Occurrence Analysis Methodology (SOAM) – A "reason"-based organizational methodology for analysing incidents and accidents. *Reliability Engineering and System Safety*, 92(9), 1162–9.

Rasmussen, J. (1997). Risk management in a dynamic society: a modeling problem. *Safety Science*, 27(2–3), 183–213.

Reason, J. (1990). *Human Error*. New York: Cambridge University Press.

Reason, J. (1991). Identifying the latent causes of aircraft accidents before and after the event. In: *Proceedings of the 22nd ISASI Annual Air Safety Seminar*, Canberra, Australia. Sterling, VA: ISASI.

Reason, J. (1997). *Managing the Risks of Organizational Accidents*. Aldershot: Ashgate.

Reason, J. (2008). *The Human Contribution: Unsafe Acts, Accidents and Heroic Recoveries*. Farnham: Ashgate.

SKYbrary. (2010). NOSS in ATM. Accessed on-line 24 October 2010 at: <http://www.skybrary.aero/index.php/NOSS_in_ATM>.

Wiegmann, D.A. and Shappell, S.A. (2003). *A Human Error Approach to Aviation Accident Analysis: The Human Factors Analysis and Classification System*. Burlington, VT: Ashgate.

World Association of Nuclear Operators. (2005). WANO review: safety first. London: Author.

Chapter 11

Safety Reporting System as a Foundation for a Safety Culture[1]

Teresa C. D'Oliveira

ISPA – Instituto Universitário, Portugal

Introduction

Major organizational accidents such as the destruction of the space shuttle *Challenger* in 1986, the explosion of the Chernobyl's nuclear power plant in 1987, the accident with off-shore platform Piper Alpha in 1988 or the destruction of the space shuttle *Columbia* in 2003, highlighted the relevance of human contributions to organizational safety. Investigations traditionally considered technical and human factors in the development and prevention of these negative events but, in spite of such operational perspective, statistics have revealed the preponderance of human factors in up to 60–70 percent of the situations (e.g., Dekker, 2002).

Applied and academic efforts strived to identify the specific behaviors that could be associated with these negative outcomes in order to prevent them.

A first step in such endeavor was to understand what human error is. Models proposed by Rasmussen (1983, 1990), Reason (1990) and Hollnagel (1998) on human error all seem to highlight a common feature: human error is presented as a category of human behaviors (D'Oliveira, 2006).

The specific behaviors that can be identified vary according to the model considered but, in general, all proposals emphasize malfunctions or deficiencies in human information processing.

Reason's approach to human frailties is probably the most well-known approach with two major categories of behaviors being considered: unintentional errors such as slips, lapses, mistakes and the noncompliance with work rules and procedures (Reason, 1990, 1998).

An additional common facet seems to be present in all models: performance is evaluated at the individual level, that is, organizational safety is associated with inadequate individual behavior.

More recently the relevance of contextual factors such as the immediate social environment and how it may trigger inadequate human action has been analyzed (e.g., Reason, 1998).

---

1 An initial version of this chapter was first presented by Teresa C. D'Oliveira and Alexandra Franco at the 28th Conference of the European Association for Aviation Psychology, Valencia, Spain, October 27–31, 2008.

Kirwan (1999) considers that there are some situations that work as catalyzers to inadequate human intervention namely the non familiarity with the task, time pressures, and power operator-system interface among others as illustrated in Table 11.1.

Table 11.1   Error producing conditions and human reliability (adapted from Kirwan, 1999)

Error producing condition	Risk factor
1. Unfamiliarity with a situation	× 17
2. Little time to detect and correct errors	× 11
3. Low signal-to-noise ratio (i.e., information deficit)	× 10
4. Poor operator-system interface	× 8
5. Mismatch between the operator's and the designer's model	× 8
6. No means to reverse unintended actions	× 8
7. Information overload	× 6
8. Inadequate risk perception	× 4
9. Poor, ambiguous or inadequate system feedback	× 4
10. Operator's inexperience	× 3
11. Mismatch between educational achievements of the operator and the task requirements	× 2
12. Incentives to use other more dangerous procedures	× 2

Reason (1990) distinguished errors in terms of their consequences. Active errors are those "whose effects are felt immediately" (p. 173) while latent errors regard adverse consequences that could be hidden for a long time in the system and only become visible when combined with other factors. While the active or manifest errors are usually associated with operational work, latent errors are related to activities "removed both in time and space from the direct control interface" (p. 173) such as designers, maintenance, high-level decision makers and managers. In most accidents, operators inherit a system with frailties created by poor design, inappropriate installation, flawed maintenance and bad decision making.

Ramanujam (2003) argues that important organizational disasters have identified organizational procedures and policies, the latent factors or characteristics, as essential precursors of adverse organizational consequences in a variety of settings such as aviation accidents, chemical disasters, nuclear power plant accidents, product recalls in the automobile industry, and losses in financial institutions or deaths from medical errors.

Reason (1998) considers that organizational disasters may be associated with different organizational deficiencies which may be traced to the mission and strategic options of a company. Various authors claim that some organizational characteristics and procedures (or the lack of them) can be viewed as "an accident in waiting".

D'Oliveira (2006), while reviewing the literature on organizational safety, identified different levels of influence (see Figure 11.1).

Top-level decisions regarding organizational objectives, the dominant values in terms of organizational culture and the way resources are managed can be considered as the most general and macro influences on organizational safety. Decisions at this level may create deficiencies in a variety of practices such as supervision, training, planning and performance monitoring. In many contexts, accident investigation is conducted in order to satisfy legal obligations but is not truly considered an opportunity for organizational learning. These organizational practices can then be translated into unsafe work conditions and processes, which in turn act as catalyzers of inadequate human intervention, that is, errors and noncompliance with rules and work procedures.

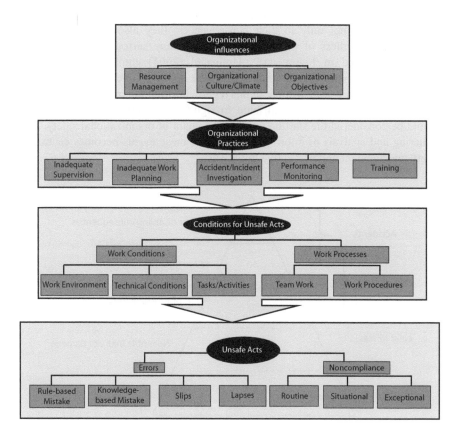

Figure 11.1   Latent catalysers of organizational safety (D'Oliveira, 2006)

The adoption of a system perspective prompted Hollnagel (2004) to identify system conditions that regard the ineffective protection of the organization. Three major latent conditions can be considered: design barriers (e.g., procedures associated with an activity) are missing or are dysfunctional, resources to minimize or neutralize potential threats are insufficient and the highly precarious functioning of the system amplifies any small active failure.

What one has to consider is that current economic conditions can actually trigger these latent conditions and factors. For example, the economic constraints companies are facing and the cost-cutting policies adopted by top management may create the background for organizational practices with consequences deferred in space and time, "an accident in waiting" as previously proposed.

The acknowledgement of safety as a social construction prompted the development of proactive approaches. If most safety elements are embedded in the organization, then one should look for their presence instead of analyzing their consequences. In this regard, Hollnagel (2004) considers that adverse consequences may arise in many forms and degrees of severity and proposed a pyramid of failures and their consequences as illustrated in Figure 11.2.

Events differ not only in their consequences; they also vary in terms of proportion. Regardless of the controversies that may surround the proportion numbers indicated in Figure 11.2, Hollnagel (2004) emphasizes that these events represent symptoms of what needs to be improved and, as such, learning opportunities for the organization. In this regard, information on the most frequent but less costly events may contribute to significant safety improvements and should be considered the starting point of the chain of organizational safety. The gathering and analysis of information regarding these events becomes central to safety management and preventive initiatives.

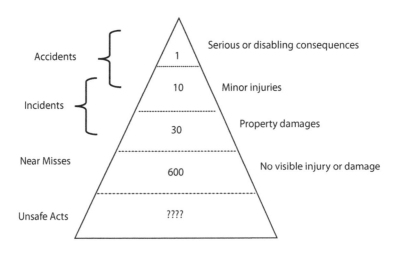

Figure 11.2   The pyramid of failures (adapted from Hollnagel, 2004)

In other words, the use of event reporting programs became central to the development of proactive approaches to aviation safety.

At the same time, a change in approach to organizational safety can be identified: a more traditional reactive perspective to undesirable events has progressively been replaced by a proactive outlook that contemplates a set of elements in an integrated system. Within this framework, latent errors are presented as a recurring theme in theoretical accounts of major accidents and, therefore, should be targeted by preventive initiatives.

The centrality of information for safety management was also highlighted by Reason (1998). According to the author an effective safety culture requires five main pillars:

1. The information element, that is, the organization has an information system where data regarding real and potential adverse events is collected, analyzed and disseminated; the system also considers data from the regular checks.
2. The reporting component where any human element of the system feels the need to participate and provide information regarding their own behavior, the behavior of colleagues or the events that took place.
3. The incentive factor whereby people perceive an atmosphere of trust and feel comfortable in reporting events. This is should not be viewed as synonymous with a blanket amnesty since it would contribute to a lack of credibility of the system.
4. A flexible organizational structure is considered a central feature of a crisis-prepared organization as learning can only take place when the organization accepts change.
5. Finally, organizations must be willing to introduce changes in their regular practices in order to accommodate recommendations from their safety information system.

Harper and Helmreich (2003) consider that the use of reporting programs can be traced to the development of the Aviation Safety Reporting System (ASRS) and more recently, has been further encouraged by the Aviation Safety Action Program (ASAP).

Two major goals were setup for the ASAPs:

1. To gather information regarding safety issues and provide protection to its proponent.
2. To use this information to develop corrective actions to reduce the potential for recurrence of accidents, incidents or safety related problems (Harper and Helmreich, 2003).

The objective of this chapter is to summarize the relevance of a reporting culture in high-reliability industries and, in particular, in the aviation industry. The chapter considers the literature on reporting programs; the safety reporting program developed at a Portuguese commercial airline, the initial results and suggested interventions.

Method

Participants

Similarly to the ASRS, the airline decided to launch a reporting safety system that included a human factors component. The Human Factors Report was initially launched in June 2006 and was made available to all personnel at the company. Anyone willing to report a safety event was welcome to do so and simply had to fill in the form that was available at the Crew Terminal. Participation was anonymous but volunteers could identify themselves if thought relevant. A total of 214 reports were received between June 2006 and August 2008 and were considered for the analysis.

Instrument

A reporting form was elaborated upon the main recommendations of the literature. A first section of the form considered personal data such as job position, and aircraft involved, activities under way when the event took place such as flight phase and cabin crew activities. A second part of the form regards the Human Factors issues involved in the event. IATA's categories of human factors to be considered in safety events were adopted (IATA, 2006 – Table 11.2). A third part of the form considers the organizational factors involved in the occurrence, the type of event and the characteristics that affected the quality of human performance.

Table 11.2   Human factors categories involved in the event (IATA, 2005)

Intentional non compliance
Deliberate and premeditated deviation from operator procedures and/or regulations. Examples include intentional disregard of operational limitations or SOPs.
Proficiency
Crew performance failures due to deficient knowledge or skills. This may be exacerbated by lack of experience, knowledge or training. Examples include inappropriate handling of the equipment.
Communication
Miscommunication, misinterpretation or failure to communicate pertinent information within the crew or between the flight crew and an external agent. CRM issues typically fall under this category. Examples include failures in monitoring/cross-checking, misunderstanding an instruction.
Procedural
Unintentional deviation in the execution of operator procedures and/or regulations. The crew has the necessary knowledge and skills, the intention is correct, but the execution is flawed. It may also include situations where crews forget or omit relevant appropriate action.
Incapacitation/Fatigue
Crew member unable to perform duties due to physical or psychological impairment.

Results

An initial analysis of the results suggest that there is a growing interest in the program with an increasing number of participants reporting events in which they were personally involved (29.4 percent in 2006; 45 percent in 2007 and 60.3 percent in 2008). Table 11.3 summarizes data regarding the operator involved, the fleet and phase of the flight when the event in question took place. Events reported involved mainly medium range flights (narrow body fleet) and suggest that the central role in the event shifted from pilots (62.5 percent in 2006) to cabin crew (77.2 percent for both cabin crew and senior cabin crew). As for the flight phase in which the event took place, a pattern of results appears to emerge as in 2006 the majority of occurrences took place outside the flying phase (dispatch, pre-flight, park) and in 2007 and 2008 during the flight itself (during take-off, cruising or descending).

Table 11.3   Main characteristics of the events reported (values in percentages)

Operator involved	2006	2007	2008
Captain	62.5	14.5	15.8
First Officer	0.0	10.1	5.4
Supervisor	0.0	4.3	1.5
Senior Cabin Crew	12.5	17.4	19.8
Cabin Crew	25.0	53.6	57.4
Fleet			
Narrow Body	71.4	76.5	75.9
Wide Body	28.6	23.5	24.1
Flight Phase			
Dispatch	12.5	2.0	2.7
Pre-flight	12.5	2.0	3.6
Pushback	0.0	7.8	4.0
Taxi-out	0.0	3.9	3.6
Take-off	0.0	33.3	5.8
Climb	0.0	5.9	8.5
Cruise	6.3	27.5	17.0
Descent	6.3	9.8	11.2
Holding	0.0	3.9	2.7
Approach	6.3	7.8	5.4
Landing	6.3	3.9	4.9
Taxi-in	0.0	0.0	0.0
Park	25.0	13.7	7.6
Various	25.0	2.0	22.9

Results regarding potential human factors involved in the event (Figure 11.3) suggest that it is possible to identify two major trends in the reports analyzed. Proficiency, communication and intentional noncompliance had a negative evolution from to 2006 to 2008. In contrast, incapacitation/fatigue and procedural issues increased in the period considered. While incapacitation or fatigue refers to the physical or psychological impairment of crew members, procedural regards slips and lapses in the execution of procedures and/or regulations.

In what regards the organizational factors involved in the event, both planning and human resources issues appear to be involved in the occurrences described with the former being responsible for the large majority of reports. This result contrasts with the consistent decreases in reports identifying training, organization and other categories as contributing factors (see Figure 11.4).

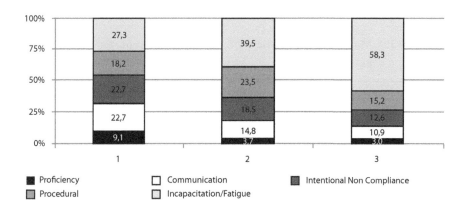

Figure 11.3   Human factors involved in the events reported

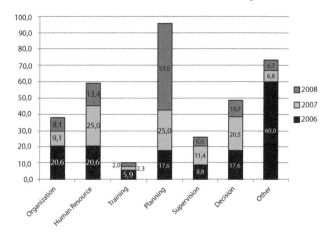

Figure 11.4   Organizational factors involved in the events reported

Discussion

The objective of this chapter is to review the relevance of a reporting culture in high-reliability organizations and, in particular, in the aviation industry. The chapter summarizes the main safety models on human intervention and highlights the relevance of collecting and analyzing data on safety occurrences. The central role of safety reporting programs for the development of an effective safety culture was also considered.

Initial data of a safety reporting program developed at a Portuguese commercial airline were analyzed. Adherence to the program is emphasized by the growing number of reports being submitted.

Results suggest that participants from medium range flights report being involved in more safety events, an involvement that is mainly linked with fatigue or incapacitation of crew members – issues that they associate with planning, human resources and organizational factors. In particular, crew members refer to rostering issues and how their work schedules contribute to higher levels of fatigue, a complaint that reflects current performance concerns in the industry (e.g., D'Oliveira, 2011).

If one is to adopt a more proactive perspective regarding safety then general recommendations should be put forward regarding a minimum number of days off to act as a counter measure for fatigue, a guideline that already exists for the long range flights. Current discussions regarding flight duty times and fatigue risk management systems also reflect these concerns.

The benefits of implementing a safety reporting system appear to be emphasized in this study as results suggested priorities of intervention in what concerns the operators' perspective. The adherence of operators and their genuine participation in such initiatives is not always problem free, especially if a blame culture exists and trust is not central in human interactions. The growing number of safety reports submitted suggests that crew members may view this initiative as a voice mechanism within the airline. It is recommended that main results of the program should be validated by crew members and that the airline communicates the applied interventions that were setup as a consequence. As with any reporting system, if no visible consequences emerge from the data collected and no clear information is conveyed to organizational members, gradual apathy and lack of interest may replace the motivational drive that was identified in implementation of this program. It is up to the airline to respond to this global adherence of operational participants and to avoid it becoming an inconsequential organizational safety ritual.

References

Dekker, S. (2002). *The Field Guide to Human Error Investigations*. Aldershot: Ashgate.

D'Oliveira, T. (2006). Desastres de origem humana: acidentes organizacionais e factores humanos [Disasters with human origin: organizational accidents and human factors]. In: M.P. Cunha and J. Gomes (eds), *Comportamento Organizacional e Gestão. 21 Temas e Debates para o Século XXI* (pp. 61–76). Lisboa: RH Editora.

D'Oliveira, T. (2011). Occupational fatigue: implications for aviation. In: K.W. Kallus and M. Heese (eds), *Aviation Psychology in Austria 2* (pp. 51–59). Vienna: facultas.wuv universitatsverlag.

Harper, M.L. and Helmreich, R. (2003). Creating and maintaining a reporting culture. Paper presented at the 12th International Symposium on Aviation Psychology, 14–17 April, Dayton, Ohio, USA.

Hollnagel, E. (1998). *Cognitive Reliability and Error Analysis Method: CREAM.* Oxford, UK: Elsevier Science.

Hollnagel, E. (2004). *Barriers and Accident Prevention.* Aldershot: Ashgate.

IATA (2006). *Safety Report 2005.* Genève: IATA.

Kirwan, B. (1999). Human reliability assessment. In: J.R. Wilson and E.N. Corlett (eds), *Evaluation of Human Work: A Practical Ergonomics Methodology* (pp. 921–68). London: Taylor and Francis.

Ramanujam, R. (2003). The effects of discontinuous change on latent errors in organizations: the moderating role of risk. *Academy of Management Journal,* 46(5), 608–17.

Rasmussen, J. (1983). Skills, rules, knowledge: signals, signs and symbols and other distinctions in human performance models. *IEEE Transactions: Systems, Man and Cybernetics,* SMC-13, 257–67.

Rasmussen, J. (1990). The role of error in organizing behavior. *Ergonomics,* 33, 1185–99.

Reason, J. (1990). *Human Error.* Cambridge: Cambridge University Press.

Reason, J. (1998). *Managing the Risks of Organizational Accidents.* Aldershot: Ashgate.

Chapter 12
Conclusions: Extending the Chain

Teresa C. D'Oliveira

ISPA – Instituto Universitário, Portugal

Previously, most references on Human Factors in Aviation have considered such traditional topics as basic psychological processes (i.e., workload, stress, decision making and situation awareness), organizational practices (i.e., personnel selection and training) and technological challenges resulting from growing automation or increased airspace capacity. Improvements in system safety and performance are typical criterion of reference for a variety of interventions.

However, the isolated consideration of each topic fails to capture the dynamic nature of the aviation industry and its operational challenges. "Mechanisms in the chain of safety" was devised with two objectives: to present the most recent research and operational efforts in aviation safety and performance and to illustrate the need for full circle approaches.

In this volume, the adoption of an IPO approach (Inputs-Processes-Outputs) to safety and performance led to the identification of three major topics: inputs, coping and control mechanisms.

Inputs in the aviation industry typically involve personnel selection, the validation of selection batteries and the use of psychological measures that take advantage of recent technological developments. Distinct contributions associated with inputs mechanisms were included in this volume: proposals that take advantage of the use of computerized tasks and suggestions linked with the identification and study of new performance indicators. Potential improvements resulting from technological developments were presented by Oubaid, Zinn and Gundert and by Uyttendaele and de Voogt.

At the center of Oubaid, Zinn and Gundert's proposal are the interpersonal interactions in aviation. Programs and interventions promoting teamwork have been underlined for almost three decades by programs such as CRM—Crew Resource Management in pilots or TRM—Team Resource Management for air traffic controllers, central for basic and recurrent training. Although some form of evaluation of characteristics promoting or associated with team work is included in most selection programs, improvements are needed in the scenarios presented and the evaluations conducted. The authors propose a new computerized tool that allows multi-observations of several applicants that interact face-to-face and through their touch screens. Conventionally, the aviation industry viewed interpersonal competence as trainable but Oubaid, Zinn and Gundert propose that the inclusion of more structured selection systems may benefit training interventions, a proposal that echoes trainability concerns.

Uyttendaele and de Voogt consider that the ability to remember and execute specific delayed tasks without being prompted to do it, that is, prospective memory, is central for air traffic controllers and also propose its inclusion in air traffic controllers' selection processes. Similarly, the suggestion follows the use of computerized tasks that try to tap distinct psychological contributions.

A final proposal, presented by Oprins, Burggraaff and Roe, explores the benefits of considering individual learning curves in the evaluation of on-the-job training. The authors starting point is the limitations associated with pass-fail decisions in most training programs; such decisions are based on experts' evaluations of continuous progression. Their proposal is that such subjective appreciation may be improved by modeling learning processes. Despite the recommendations for reliability improvements of some indicators, the authors propose that the use of learning curves provides greater insight to individual learning processes, captures the dynamic nature of OTJ training and allows adaptive training. Similarly to Oubaid, Zinn and Gundert; Oprins, Burggraaff and Roe also purport that a trainability approach for selection purposes may be possible and efforts should be developed in this direction.

Common to the three proposals is the creative use of computerized measures to tap specific psychological processes and distinct performance indicators and the advantages that may be introduced in the system by exploring the links between personnel selection and training.

Coping mechanisms refer to individual and organizational processes that may better prepare one to deal with high situational demands and critical or exceptional events. Haeusler, Hermann, Bienefeld and Semmer suggest that adaptive expertise needs to be fostered in training. The adaptation to task demands and situational changes, along with the anticipation of future developments, requires situation awareness and explicit planning; in other words, it entails working smarter and not simply working harder. It is through diversified training scenarios that the ability to proactively adapt to high demands may be developed. In a clear association with different organizational factors and interventions, the authors advocate that a change in most operational philosophies is needed; one has to overcome a training perspective that emphasizes minimum compliance in favor of an approach that allows exploring the true potential of training investments.

The relevance of anticipations processes is also pondered by Kallus who recommends that the anticipation of and coping with critical flight situations must be considered in training and opportunities for the development of these skills. The author goes even further and believes that the consequences of anticipation processes need to be taken into consideration in accident-incident investigations and can also be explored in selection processes.

Identical concerns are expressed by Ebbatson, Harris, Huddlestone and Sears regarding events and situations that require manual flight. Although manual flight is limited in normal operations, recurrent training and proficiency checks require it. Since only infrequent opportunities to exercise manual flying skills exist, the authors propose that new performance indicators and training programs should be restructured to compensate.

Training issues and customized solutions are also submitted by Cherng, Shiu and Wen. Their study suggests that different stressor profiles and coping behaviors are found in Taiwanese and foreign pilots, a scenario that may be typical of multicultural airlines.

While inputs components are mainly concerned with providing the system with new information, either of psychological processes or on individual performance, coping mechanisms emphasize the dynamic nature of operations and how organizational practices may foster proactive behaviors.

Control mechanisms also stress the need for a culture of information but enlarge the concept proposed by Reason (1998). While Reason's approach refers to a profound knowledge of the organization, control interventions take into account the interconnected nature of activities and information in the system. For example, information regarding interpersonal competence, as proposed by Oubaid, Zinn and Gundert, may be relevant not only for selection purposes but also for the development of diversified training scenarios and learning processes, accident-incident investigation, among others.

The feedback loop proposed in control mechanism suggests the relevance of adaptive and customized organizational interventions.

The associations of distinct detection mechanisms to different types of errors led Thomas to propose an organic approach according to which variability in performance and human error requires parallel defense strategies and an adaptive approach and not the traditional serial protective layers as illustrated by Reason's Swiss Cheese Model.

Training implications are proposed both by Stánski-Pacis and de Voogt and Hayward, Lowe and Brandford. While the former considers technological innovations and improvements in user interface, the latter focuses on proactive training practices.

Stánski-Pacis and de Voogt's analysis of the role of GPS in aviation incidents and accidents proposes customized training according to user's profile.

Hayward, Lowe and Brandford suggest a proactive training methodology that draws on participants' experiences and that may help to promote organizational learning and safety risk management.

Finally, D'Oliveira reviewed data from a safety reporting system and highlights the need for translating the results of this voice mechanism into operational practices; the growing interest and adherence to the reporting system can be associated with a culture of information, reporting and trust. However, if the concerns expressed cannot be linked with significant organizational changes, the chain of safety is broken.

Inputs and coping mechanisms emphasize the need for information associated with system's monitoring and critical events or situations and the ability to deal with such occurrences. Control mechanisms accentuate the need to give feedback to the organization so that improvements can be introduced into normal operations.

The different components in the chain of safety discussed in this volume reflect concerns that involve different organizational levels from micro-interventions that reflect basic psychological processes such as prospective memory to macro-proposals that involve using previous experiences as training tools. A multi-level approach is therefore needed for improvements in performance and systems safety to take place.

An additional characteristic can be inferred from the distinct contributions to the chain of safety: organizational learning.

Organizational learning is the process of increasing the potential for improved organizational action through knowledge and understanding (Carroll and Edmondson, 2002) and, according to Pidgeon (2010), is a key element of effective safety cultures. Organizational learning involves internal and external flows of information, that is, sharing and learning from experience and from others (Saw, Wilday and Harte, 2010) or a culture of information (Reason, 1998). Due to heterogeneity of experiences and specificity of operations (generalized airlines versus specialist airlines) customized organizational learning is proposed in the literature (Haunschild and Sullivan, 2002).

Yeo (2002) suggests that organizational learning may be linked with three learning loops. The single loop refers to learning processes that take place at the individual level and connect the individual with specific organizational norms (e.g., personal goals, abilities, competences and mental models). Single loop processes can be enhanced by improvements in input mechanisms and individual coping strategies.

Double loop processes involve higher-level performance as team or group outputs and require greater dynamics in feedback and inquiry in order to change organizational norms (Yeo, 2002). Most control mechanisms (e.g., safety reporting systems or the analysis of GPS involvement in accident and incidents) can be understood as double loop learning processes. Improvements of work practices or processes in general and operational control efforts are associated with this systemic learning.

Triple loop learning involves the overall vision of the organization, organizational goals and strategic management. Recommendations for new operational philosophies, as proposed by Thomas and D'Oliveira, emphasize the need for "feedforward" as an organizational tool (i.e., using the culture of information to anticipate future changes and not only to gain better understanding of past events) thus promoting triple loop processes or strategic learning. Strategic learning is associated with strategic control that involves macro goal definition and its translation into policies and specific plans of action (Child, 2005).

"Mechanisms in the chain of safety" illustrates different organizational learning opportunities and the potentially rich information they can derive.

However, feedforward or triple loop processes have to be associated with operational changes. The anticipation of future events and situations and the scenario planning associated with these macro processes require organizational flexibility and play an important role in strategic learning.

If one considers that mechanisms in the chain of safety involve monitoring normal and critical events, the ability to anticipate future disruptions and the mitigation of its consequences, the capacity to learn from experience and the introduction of changes in procedures, routines, jobs and roles, then one is envisioning the operational pillars of resilient systems (Hollnagel, 2008). In order words, mechanisms in the chain of safety allow us to analyze and improve different paths to operational resilience. If research proposes that customized training opportunities should be put forward in order for technical and systemic learning to take place, then similar suggestions can be presented for macro interventions with customized strategic learning being associated with greater organizational resilience.

References

Carroll, J.S. and Edmondson, A.C. (2002). Leading organizational learning in healthcare. *Quality and Safety in Health Care*, 11, 51–6.

Child, J. (2005). *Organization: Contemporary Principles and Practice*. Oxford: Blackwell.

Haunschild, P.R. and Sullivan, B.N. (2002). Learning from complexity: effects of prior accidents and incidents on airlines' learning. *Administrative Science Quarterly*, 47, 609–43.

Hollnagel, E. (2008). Resilience engineering in a nutshell. In: E. Hollnagel, C.P. Nemeth and S. Dekker (eds). *Remaining Sensitive to the Possibility of Failure* (pp. xi–xiv). Aldershot: Ashgate.

Pidgeon, N. (2010). Systems thinking, culture of reliability and safety. *Civil Engineering and Environmental Systems*, 27, 211–17.

Reason, J. (1998). *Managing the Risks of Organizational Accidents*. Aldershot: Ashgate.

Saw, J.L., Wilday, J. and Harte, H. (2010). Learning organizations for major hazards and the role of the regulator. *Process Safety and Environmental Protection*, 88, 236–42.

Yeo, R. (2002). From individual to team learning: practical perspectives on the learning organization. *Team Performance Management: An International Journal*, 8, 157–70.

Author Index

Subject Index